Proceedings in Engineering Mechanics

Research, Technology and Education

This book series publishes the results of meetings dealing with material properties in engineering and science. It covers a wide range of topics, from the fundamentals of materials mechanics and applications for various industries to aspects of scientific training and career development. The volumes in the series are based typically on primary research materials presented at conferences, workshops, and similar scientific meetings, and represent comprehensive scientific and technical studies.

Lucas F. M. da Silva · António J. M. Ferreira
Editors

3rd International Conference on Science and Technology Education 2022

Selected Contributions of STE 2022

Springer

Editors
Lucas F. M. da Silva
Faculty of Engineering
University of Porto
Porto, Portugal

António J. M. Ferreira
Faculty of Engineering
University of Porto
Porto, Portugal

ISSN 2731-0221 ISSN 2731-023X (electronic)
Proceedings in Engineering Mechanics
ISBN 978-3-031-25400-0 ISBN 978-3-031-25401-7 (eBook)
https://doi.org/10.1007/978-3-031-25401-7

This Springer imprint is published by the registered company Springer Nature Switzerland AG
The registered company address is: Gewerbestrasse 11, 6330 Cham, Switzerland

Preface

This volume of the series *Proceedings in Engineering Mechanics—Research, Technology and Education* contains selected papers presented at the 3rd International Conference on Science and Technology Education STE 2022, held at Faculty of Engineering of the University of Porto (FEUP), Portugal, during 6–7 October 2022.

This conference is held every two years. The conference is co-chaired by António Ferreira (University of Porto, Portugal) and Lucas F. M. da Silva (University of Porto, Portugal). The goal of the conference is to provide an international forum for the sharing, dissemination and discussion of research, experience and perspectives across a wide range of teaching and learning issues. About 55 papers were presented by researchers from nearly 20 countries.

In order to disseminate the work presented at STE 2022, selected papers were prepared which resulted in the present book. Various topics are covered resulting in 10 chapters dealing with learning mechanisms (first seven papers), learning systems (one paper) and assessment (last two papers). The papers presented here are good examples of the latest trends related to science and technology education.

The organizers and editors wish to thank all the authors for their participation and cooperation, which made this volume possible. Finally, I would like to thank the team of Springer-Verlag, especially Dr. Christoph Baumann and Ute Heuser, for their excellent cooperation during the preparation of this volume.

Porto, Portugal
November 2022

Lucas F. M. da Silva
lucas@fe.up.pt

António J. M. Ferreira
ferreira@fe.up.pt

Contents

Learning Systems

Assessment

Learning Mechanisms

Influence of COVID-19 Confinement and Pandemic on the Academic Performance of Students

I. Tort-Ausina, J. Molina-Mateo, A. Vidaurre, J. M. Meseguer-Dueñas, J. Riera, J. A. Gómez-Tejedor, María-Antonia Serrano, and S. Quiles

Abstract The outbreak of the pandemic caused by the SARS-CoV-2 virus in 2020 and the consequent confinement and other restrictions imposed by the authorities posed a great challenge for the entire society, and, particularly, it was a great challenge for the educational system. To analyse the effect that the pandemic has had on students' performance, this paper analyses the academic results of three different courses at the Universitat Politècnica de València (UPV): before the pandemic (2018–19 academic year); at the beginning of the pandemic, where the strictest confinement measures were imposed (2019–20 academic year) and in later stages of the pandemic where the restrictions were eased (2020–21 academic year). We analysed three first-year physics courses from three different degrees. Despite the efforts made by teachers to try to achieve an evaluation as similar as possible to the situation before the pandemic, an apparent increase in the average grade was observed in

I. Tort-Ausina (✉) · J. Molina-Mateo · A. Vidaurre · J. M. Meseguer-Dueñas · J. Riera · J. A. Gómez-Tejedor · M.-A. Serrano · S. Quiles
Applied Physics Department, Universitat Politècnica de València. C/ Camino de Vera, s/n, 46022 Valencia, Spain
e-mail: isatort@fis.upv.es

J. Molina-Mateo
e-mail: jmmateo@fis.upv.es

A. Vidaurre
e-mail: vidaurre@fis.upv.es

J. M. Meseguer-Dueñas
e-mail: jmmesegu@fis.upv.es

J. Riera
e-mail: jriera@fis.upv.es

J. A. Gómez-Tejedor
e-mail: jogomez@fis.upv.es

M.-A. Serrano
e-mail: mserranj@fis.upv.es

S. Quiles
e-mail: suquica@fis.upv.es

L. F. M. da Silva and A. J. M. Ferreira (eds.), *3rd International Conference on Science and Technology Education 2022*, Proceedings in Engineering Mechanics,
https://doi.org/10.1007/978-3-031-25401-7_1

the year of confinement. This increase can be explained because the evaluation was fundamentally online. Students were evaluated based on academic work and online exams, where they had at their disposal several means to help them to take those exams. After confinement, the academic results before the pandemic were recovered. This fact is very remarkable since it must be considered that these students are the ones who the previous year had completed a large part of the high school course in confinement conditions. In addition, they have followed the course in a semi-classroom modality recovering the face-to-face assessment. The fact that they had academic results that were similar to those before the pandemic indicates that the course's confinement and semi-attendance have not significantly affected their subsequent academic performance. The percentage of students who withdraw from the three courses has also been analysed.

Keywords Pandemic · Students' performance · Physics · Engineering degrees · Students' withdrawal

1 Introduction

The outbreak of the pandemic caused by the SARS-CoV-2 virus in 2020 produced a drastic impact on the entire society. At the academic level, the way teaching and learning was affected at the major components: teaching methods, course content, assessment strategies, faculty's readiness and technical issues (Farnell et al. 2021; Pandya et al. 2022; Pokhrel and Chhetri 2021). Teacher made a significant effort trying to soften the impact; Pandya et al. found differences in the teaching methods and in assessment strategies, but no differences were found in the course content and in technical support and training to use educational technologies (Pandya et al. 2022).

Innovative technologies and online teaching have been vital in achieving quality teaching in this situation, thus overcoming the restrictions due to the pandemic. In a recent study performed using data from Chinese Middle Schools (Clark et al. 2021), it was found that receiving online education during the COVID-19 lockdown improved student academic results, relative to pupils without learning support from their school. But not all online education was equal: students who were given recorded online lessons from external higher-quality teachers had higher exam scores than those whose lessons were recorded by teachers from their own school. The exam performance of students who used a computer for online education was better than those who used a smartphone, and it was low achievers who benefited most from teacher quality (Clark et al. 2021). Some authors (Evans et al. 2020; Romero-Hall and Vicentini 2017) pointed out that the lack of integrated e-learning platforms and technologies had a negative impact on the online teaching environment. Some others concluded that the key obstacles that higher education institutions faced in transitioning to online courses were technical resources and differential access to education so improving digital infrastructure and reducing the cost of internet access may be necessary for mitigating the impact of the COVID-19 pandemic on education outcomes (Chisadza et al. 2021; Treve 2021). Coman et al. (2020) studied how Romanian Universities dealt with the transition to e-learning during COVID-19 and found that instructors had issues with adapting content to the online format while students presented decreased motivation and learning obstacles. Yasmin (2022) collected the opinion of learners and teachers of chemical engineering in Pakistan. The results showed high dissatisfaction among students and teachers. Students had low motivation due to inadequate electronic devices and poor connectivity. Teachers claimed institutional, social, technical and learning-management-related barriers.

Researchers are not unanimous about students' output in online courses. Romero-Hall and Vicentini (Romero-Hall and Vicentini 2017) found that students who followed the online courses could promote learners' autonomy and improve their study habits. AlMahdawi et al. (2021) found a positive impact on critical thinking, collaborative skills, creativity and innovation, technology application, class participation, and overall achievement during online and distance learning in a chemistry course during the COVID-19 Pandemic. Ripoll et al. (2021) proposed a methodology based on cooperative learning in response to the COVID-19 crisis. As a result, increased student participation in relation to previous years, and improved student-learning outcomes. Delgado (2021) studied the implementation of a digital-learning physics course for computer science students. Their results showed a positive evaluation of the digital strategy. The study indicated differences in the students' performances in the different courses with no significant differences among genders.

In this work we analyse the academic results of first-year physics courses from three different degrees at three different academic years at the Universitat Politècnica de València (UPV): before the pandemic (2018–19 academic year); at the beginning of the pandemic, where the strictest confinement measures were imposed (2019–20 academic year) and in later stages of the pandemic where the restrictions were eased. Academic grades and number of students who did not attend all the exams, considering a gender perspective, were analysed for the three consecutive academic years.

2 Methodology

The description of the basic characteristics of the three first-year courses studied are shown in Table 1.

Table 2 shows how the teaching and evaluation was adapted in the two years affected by the confinement conditions: in the second semester of the 2019–2020 academic year, and safety limitations during the 2020–2021 academic year due to the COVID-19 pandemic. During the 2020–2021 academic year, the UPV proposed three possible scenarios: in-person (face to face, F2F) teaching in those groups and classrooms where the social security distance required by health standards could be maintained, the virtual (V) teaching, totally online in cases of vulnerability of the

Table 1 General description of the courses

Course	Electricity	Physics	Physics
Degree	Electronics and Industrial Automatics (EIA)	Aerospace Engineering (AE)	Technical Architecture (TA)
Credits	6	12	4.5
Semester	2nd	1st and 2nd	2nd
Number of students	150–160 students, 3 groups	120–130 students, 2 groups	120–130 students, 3 groups
Methodology	All groups of each course are taught using the same methodology and common learning material and evaluation The methodology is active/constructive, enhancing the student participation in problem solving Written material and videos of the content are available to students. Some material is intended for students' autonomous use, and some material is to be used during the classes		
Evaluation	The final grade (FG) includes several evaluation items: • Individual work of the students, through 2 written exams and 6 online exams: 65% of the FG • Teamwork in the laboratory practices and in the solved problems and their presentation to the group: 35% of the FG	The final grade (FG) includes several evaluation items: • Individual work of the students, through 4 written exams: 70% of the FG • Team work in the laboratory practices: 20% of the FG • Individual and teamwork performed daily during the classes: 10% of the FG	The final grade (FG) includes several evaluation items: • Individual work of the students, through 2 written exams: 75% of the FG • Teamwork in the laboratory practices and in the oral presentation: 25% of the FG

teacher, and the hybrid (Synchronous Hybrid Learning, SHL) teaching in the rest of the cases, with half of the students following in person the teaching in the classroom, while the rest of the students were following the class in streaming, outside the classroom.

3 Results

This study analyses and compares the effect of the pandemic for the three before mentioned courses. This is the reason why three academic years are studied: 2018–2019 (pre-pandemic), 2019–2020 (confinement), 2020–2021 (pandemic).

Statistical analysis has been performed with SPSS, with a significance level value of $\alpha = 0.05$ for the different tests.

Table 2 Adaptation during academic years affected by COVID-19

<table>
<tr><th></th><th>Electricity
Electronics and Industrial Automatics (EIA)</th><th>Physics
Aerospace Engineering (AE)</th><th>Physics
Technical Architecture (TA)</th></tr>
<tr><td rowspan="5">2019–2020</td><td>152 students</td><td>122 students</td><td>126 students</td></tr>
<tr><td colspan="3">Since the beginning of the confinement due to COVID-2019 (March 2020), the entire course was taught online. The commitment to face the new situation, both by teachers and students, allowed to follow the same methodology, thanks to the use of Microsoft® Teams. The classes were recorded and were available online for the students</td></tr>
<tr><td colspan="3">The written exams were individual, generated through the application of the university platform, in which, in addition to automatic correction, they had to upload the written document to solve the exam review requests</td></tr>
<tr><td colspan="3">The laboratory practices were adapted by using demonstrative videos, providing the students with some experimental measurements (which they could not acquire in the laboratory), and in some cases substituting the measurements in the laboratory by measurements taken at home using an App for the cellular phone</td></tr>
<tr><td colspan="3">Tutorials were carried out by email and Teams in all the groups</td></tr>
<tr><td rowspan="4">2020–2021</td><td>155 students</td><td>124 students</td><td>119 students</td></tr>
<tr><td>One group followed a face-to-face method with 22 students (F2F); another online group with 69 students (V), and the third with hybrid methodology with 64 students (SHL). All three groups had access to the same resources. In groups V and SHL the classes were broadcasted in streaming. In group V they were recorded and available for students</td><td>A hybrid methodology (SHL) was applied: half of the class attended in person, while the other half did so online in alternating weeks</td><td>All groups were taught face-to-face (F2F), the same as before the pandemic, because the capacity of the classrooms allowed face-to-face learning</td></tr>
<tr><td colspan="3">Tutoring by email and Microsoft Teams was performed in all the groups, and in the F2F and SHL groups the tutoring was also face-to-face</td></tr>
<tr><td colspan="3">The written exams were carried out in person</td></tr>
</table>

3.1 Evolution of the Student Withdrawal

To start studying the possible effects of the pandemic, the students' withdrawal data were analysed. Figure 1 shows the evolution of the percentage of student withdrawal during the three years, for the three courses. The term withdrawal refers to the point at which students stop their scheduled learning, and do not attend teaching and assessment activities.

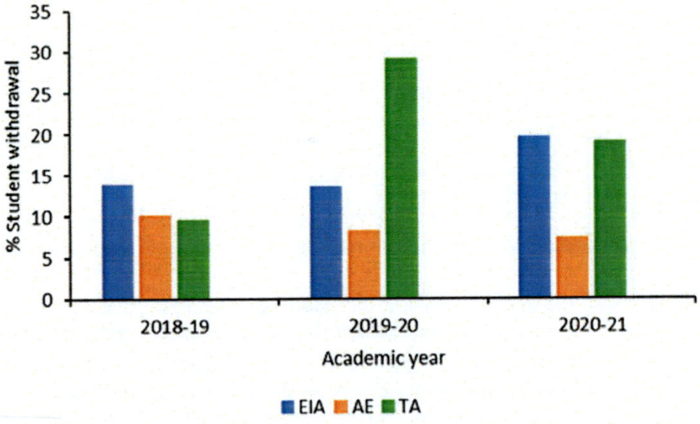

Fig. 1 Evolution of the percentage of student withdrawal for the three courses

The percentage of student withdrawal shows a great variability, both for the academic years and for the degrees. In one hand, it is worth noting the increase in the withdrawal in TA (Technical Architecture degree) during the confinement course. On the other hand, withdrawal is very high (around 20%) in the pandemic year both for TA and EIA.

In order to analyse this withdrawal from a gender perspective, Fig. 2 shows the evolution of this withdrawal for the three courses according to the student's gender.

In general terms, the number of student withdrawal is higher in males than in females. To determine whether there were statistically significant differences between

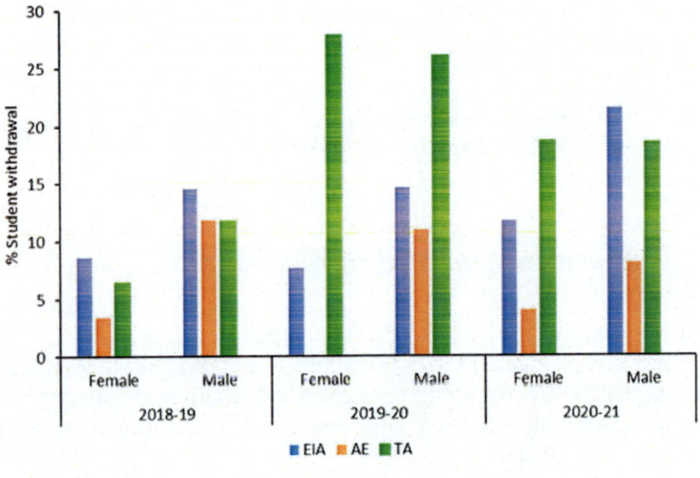

Fig. 2 Evolution of the percentage of student withdrawal among gender

the withdrawal of males and females, a Wilcoxon Signed Ranks Test was performed: the analysis revealed a statistically significant difference between the average of the percentage of withdrawal in each degree of men and those of women.

It should be noted that in the year of the pre-pandemic this difference of withdrawals (in %) is approximately double in male than in female. This trend is broken during the year of the confinement and pandemic in the TA degree, but it remains the same in the other two degrees.

3.2 *Evolution of the Final Grades*

In order to show the differences in the students' performance during the pandemic, the final grades of the three courses were analysed. Figure 3 shows the evolution of the average of final grades of all the groups of each course and their standard deviation.

In the three courses, the grades were higher during the confinement; one can assume that this increase in the grades may be due to the different type of evaluation that had to be followed during the confinement. It can also be seen how the standard deviation is of the same order of magnitude in the three years for each degree. Other authors have also found that students achieved better results under a temporary shift of instructional system to an alternate remote system due to COVID-19 circumstances (Gonzalez et al. 2020; Iglesias-Pradas et al. 2021).

As our main interest were the possible changes in the academic performance during these three years, the results of the grades of the three degrees were grouped for each year. Then, an ANOVA test was carried out in order to verify if there are significant differences in the averages of the grades between the three academic

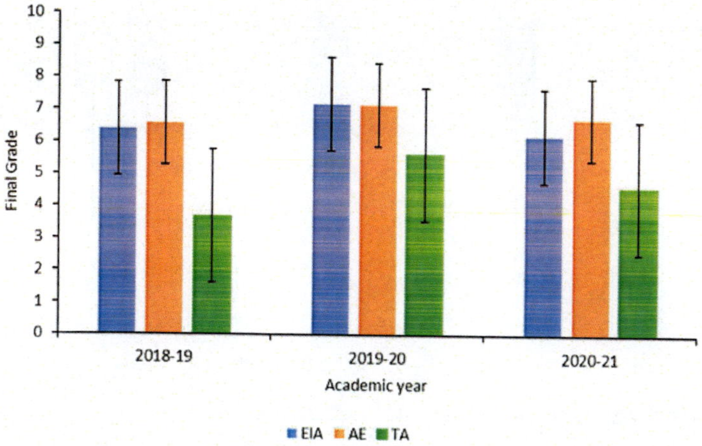

Fig. 3 Temporal study of the final grades

years. This analysis has revealed statistically a significant differencc ($p < 0.001$), so a contrast test was performed between the three academic years. The result of the contrast test ($p < 0.001$) indicates that the average grade for the 2019–20 academic year (confinement) is significantly higher than that of the other two academic years.

Carrying out the same study for each degree independently, the previous result is shown again ($p < 0.001$), being consistent with our initial expectations.

In addition, in the TA degree, the grade of the 2020–21 academic year (pandemic) is higher than that of the 2018–19 academic year (pre-pandemic) ($p < 0.001$), while in the other two degrees there are no significant differences between the grades of these two courses.

To complete this study of the differences in the students' performances in the different courses, significant differences among genders were analysed. Figure 4 shows the final grades of each year (grouping the grades of the three courses in each year) of males and females.

The analyses shows that the average final grade of males is statistically higher ($p < 0.018$) than females.

When performing the two-way ANOVA test according to year and gender to study if there are significant differences between the grades, we see in the results that there are significant differences between academic years among gender ($p < 0.01$), being that of the female students significantly lower. It is worth mentioning that in the year of the confinement the average grade is the highest and these differences are minimized.

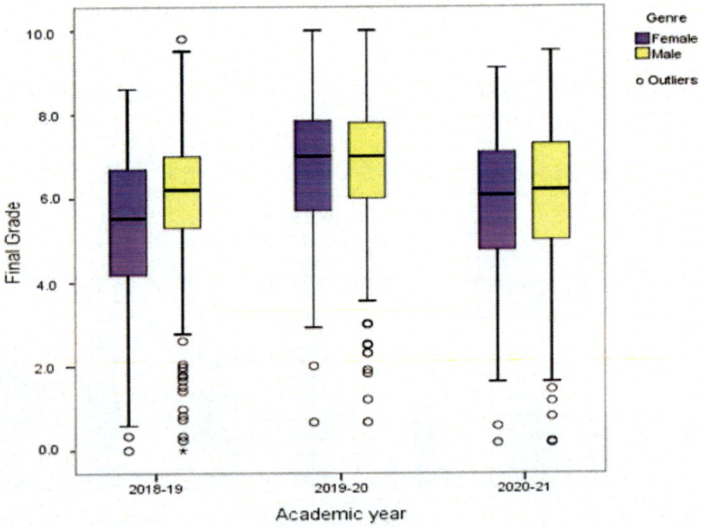

Fig. 4 Evolution of the final grades of males and females

4 Conclusions

The main conclusion in this work is that the effect of confinement and pandemics on students was extraordinarily strong, and normality has not still been reached. The principal effect could be observed in withdrawals during the course. The confinement increased dramatically the number of students that gave up the courses. This number has decreased during the pandemic period, but it has not reached the values that we had before the pandemic.

The number of withdrawals of male students was significantly higher before the pandemics, but during confinement the abandons of female students increased in such a way that both genders had a greater and equal number of withdrawals.

Grades during the confinement experienced a clear increase when compared to the previous situation. Considering the additional difficulties that students had to face during this period, this can only be explained by the changes in the evaluation system. During the period before the confinement, male students obtained better grades, but the increase in grades observed during the confinement balanced the difference between both genders.

Normality has still not been reached, and all these results show that we must pay attention to the withdrawals in our courses and possible gender gaps that could remain for next courses if we do not correct them.

References

AlMahdawi, M., Senghore, S., Ambrin, H., Belbase, S.: High school students' performance indicators in distance learning in chemistry during the Covid-19 pandemic. Educ. Sci. **11**, 672 (2021). https://doi.org/10.3390/educsci11110672

Chisadza, C., Clance, M., Mthembu, T., Nicholls, N., Yitbarek, E.: Online and face-to-face learning: evidence from students' performance during the Covid-19 pandemic. Afr. Dev. Rev. **33** (2021). https://doi.org/10.1111/1467-8268.12520

Clark, A.E., Nong, H., Zhu, H., Zhu, R.: Compensating for academic loss: online learning and student performance during the Covid-19 pandemic. China Econ. Rev. **68**, 101629 (2021). https://doi.org/10.1016/j.chieco.2021.101629

Coman, C., Țîru, L.G., Meseșan-Schmitz, L., Stanciu, C., Bularca, M.C.: Online teaching and learning in higher education during the coronavirus pandemic: students' perspective. Sustainability **12**, 1–22 (2020). https://doi.org/10.3390/su122410367

Delgado, F.: Teaching physics for computer science students in higher education during the Covid-19 pandemic: a fully internet-supported course. Future Internet **13**, 1–24 (2021). https://doi.org/10.3390/fi13020035

Evans, D.J.R., Bay, B.H., Wilson, T.D., Smith, C.F., Lachman, N., Pawlina, W.: Going virtual to support anatomy education: a STOPGAP in the midst of the Covid-19 pandemic. Anat. Sci. Educ. **13**, 279–283 (2020). https://doi.org/10.1002/ase.1963

Farnell, T., Skledar Matijevic, A., Šcukanec Schmidt, N.: The impact of COVID-19 on higher education: a review of emerging evidence. Analytical report. Publications Office of the European Union, Luxembourg (2021)

Gonzalez, T., de la Rubia, M.A., Hincz, K.P., Comas-Lopez, M., Subirats, L., Fort, S., Sacha, G.M.: Influence of Covid-19 confinement on students' performance in higher education. PLoS ONE **15**, e0239490 (2020). https://doi.org/10.1371/journal.pone.0239490

Iglesias-Pradas, S., Hernández-García, Á., Chaparro-Peláez, J., Prieto, J.L.: Emergency remote teaching and students' academic performance in higher education during the Covid-19 pandemic: a case study. Comput. Human Behav. **119**, 106713 (2021). https://doi.org/10.1016/j.chb.2021.106713

Pandya, B., Patterson, L., Cho, B.Y.: Pedagogical transitions experienced by higher education faculty members—"Pre-Covid to Covid." J. Appl. Res. High. Educ. **14**, 987–1006 (2022). https://doi.org/10.1108/JARHE-01-2021-0028

Pokhrel, S., Chhetri, R.: A literature review on impact of Covid-19 pandemic on teaching and learning. High. Educ. Future **8**, 133–141 (2021). https://doi.org/10.1177/2347631120983481

Ripoll, V., Godino-Ojer, M., Calzada, J.: Teaching chemical engineering to biotechnology students in the time of COVID-19: assessment of the adaptation to digitalization. Educ. Chem. Eng. **34**, 21–32 (2021). https://doi.org/10.1016/j.ece.2020.11.001

Romero-Hall, E., Vicentini, C.: Examining distance learners in hybrid synchronous instruction: successes and challenges. Online Learn. J. **21**, 141–157 (2017). https://doi.org/10.24059/olj.v21i4.1258

Treve, M.: What Covid-19 has introduced into education: challenges Facing Higher Education Institutions (HEIs). High. Educ. Pedagog. **6**, 212–227 (2021). https://doi.org/10.1080/23752696.2021.1951616

Yasmin, M.: Online chemical engineering education during Covid-19 pandemic: lessons learned from Pakistan. Educ. Chem. Eng. **39**, 19–30 (2022). https://doi.org/10.1016/j.ece.2022.02.002

3D with Artificial Intelligence Technology Contributes Science and Technology Education Development

Hao Jiang, Xiaoli Liu, Mingxi Tang, and Kaibin Xiang

Abstract The combination of human cognition, digitalization and artificial intelligence informed the core technologies in Industry 5.0 generation. According to description of EU commission "Industry 5.0 focuses on technology and innovation as necessary components of the transition to a new industrial paradigm". Human, objects, computing devices connect by a giant digital network under the industrial framework. Machine, human and robot will help each other breakthrough computing boundaries in this generation. This study discusses the relationship between design, and STEM education, and how 3D and artificial intelligence technology as a medium to contribute to STEM education in fifth industrial revolution.

Keywords Design technology · Collaborative learning · Creative · STEM · Design-STEM

1 Introduction

Globalization of economies, technologies, and international relationships in rapid development. Transfer to Industry 5.0, many people and companies wanted to bring the human, social, technology, and environmental into creativity, efficient and inteligent equation. The great change forced education to reconsider how teaching and

H. Jiang (✉)
H.Cruiser Informationsgesellschaft mbH, 80807 Munich, Germany
e-mail: celestehao@gmail.com

X. Liu
Shanxi Fashion Engineering Institute, Qingdu District, Xianyang, Shanxi 712099, China

M. Tang
Shenzhen University, 3688 Nanhan Road, Nanshan, Shenzhen, Guangdon 518060, China

K. Xiang
Kelu Education Science and Technology Co., Ltd, Keji Rd 8 S, Nanshan, Shenzhen, Guangdong 518063, China

L. F. M. da Silva and A. J. M. Ferreira (eds.), *3rd International Conference on Science and Technology Education 2022*, Proceedings in Engineering Mechanics,
https://doi.org/10.1007/978-3-031-25401-7_2

learning take place in the lately educational system and explore any innovative, effective approaches to schooling [1]. Since the 1990s, a new pedagogical scheme that integrated science, technology, engineering, and mathematics had been first proposed by a variety of educators in the United States, which called STEM education aims to develop students' work in the multidisciplinary industry.

There are various ways and learning tools to assist students in obtaining knowledge of STEM (science, technology, engineering, mathematics) subjects. Researchers aim to propose a diversified perspective that introduces design technology to support STEM teaching and learning in three aspects. The first is which parts of the STEM education system should be upgraded and focused on in the era of Industry 5.0. For example, which skills need to be developed? What kind of rules have to be defined? Which impact may AI have? The second we explained how the design technology, which involved 3D modeling, optical scanning, and virtual reality applied to the practice process. At last, we will compare the difference between existing methods with ours. Conclude the impact of 3D design technology on students' cognition during learning and how this improved science and technology education development.

2 Definition and Core Technology of Industry 4.0

Industry 5.0 can seem as an extension of the previous generation (Industry 4.0). Therefore, it is important to have a general understanding of Industry 4.0 and its core technology. The definition of Industry 4.0 is "a name for the existing trend of automation and data exchange in manufacturing technologies," which presents a new phase in the organization and control of the industrial value chain. It refers to the application of smart systems for the future and processes for the industry with the help of information and communication technology [2]. Under this stage, human beings are the most flexible and most intelligent components of both the present and future. Machines are becoming smart enough to be used efficiently by humans [3]. It comes from the German term "Industrie 4.0" which transformed German manufacturing with the Internet of Things and cyber-physical systems to the center stage through high-tech strategy. It also focuses on production, people, the environment, and security. Therefore, the design principles of Industry 4.0 are Interoperability, Virtualization, Decentralization, Real-time capability, Service orientation, and Modularity. In the meantime, Artificial Intelligence is the main technology that carries out most routine tasks.

By definition, AI refers to human-like intelligence exhibited by machines like natural intelligence able to help people to solve problems of various natures [4]. AI technologies are widely used in the manufacturing industry and consolidate physical and virtual worlds to assist cyber-physical systems. Therefore, AI integrates machines to complete complex tasks, decrease costs, and improve the quality of production and service is the core principle of smart manufacturing and industry 4.0 [5]. The AI algorithms help factories to optimize the product design and process, and they also can share assembly lines to increase efficiency. It is more than a tool, like a medium [6] to create meaningful perspectives that contribute to creativity in manufacturing [4].

3 Transforming from Industry 4.0 to Industry 5.0

Industry 5.0 does not like a revolution, it is more like an upgrade and refinement of industry 4.0. According to the guide of the European Commission, Industry 5.0 integrates the swerving strengths of cyber-physical production systems and human intelligence to create synergetic factories [7]. It brings benefit to industries, workers, and society. Compared with Industry 4.0, the fifth evolution focuses more on the original principles of social fairness and sustainability supports to the industry in its long-term service to humanity within planetary boundaries [8]. Applications of Industry 5.0 include intelligence healthcare, cloud manufacturing, supply chain management, manufacturing, education, and human-cyber-physical system. The relative technological trend integrated with cognitive skills and innovation involves edge computing, digital twins, collaborative robots, internet of everything, big data, 6G, and blockchain [7]. The combination of applications and technologies enables industry 5.0 to develop into an advanced manufacturing model which forces interaction between humans and computers (machines). Compared with Industry 4.0, a noticeable difference in Industry 5.0 is that machines collaborate with humans, and this feature is particularly noteworthy in the covid-19 pandemic era. Amounts of people have to change their lifestyle, which includes working, studying, and shopping from offline to online. Artificial intelligence with computing technologies has become the primary medium and functional tools to contribute people complete assignments, tasks, or preparation for daily life.

4 The Impact of Industry 5.0 on STEM Education

Industry 4.0 and 5.0 refer to a variety of industries and enabling technologies. According to the report by European Commission named "Enabling technologies for Industry 5.0", Individualized human–machine-interaction; Bio-inspired technologies and smart materials; Digital twins and simulation; Data transmission, storage, and analysis technologies; Artificial intelligence; Technologies for energy efficiency, renewables, storage, and autonomy; each of them can only achieve its potential when combined with other categories as a part of system and technology framework [8]. In summary, Industry 5.0 or Society 5.0 aims to support a solution to social problems with the help of the integration of physical and virtual spaces that would be achieved by Industry 4.0 [9]. Undoubtedly, Artificial Intelligence technology is a continuation of the previous era and plays a key role between machines and humans.

STEM education and industries should be complemented and promoted each other; a reasonable combination of them is a topic worthy of consideration. Today's educators are able to build on the vision of students' communication and collaboration skills, integrating technology and problem-solving skills, and improving innovative ability and creative thinking [10]. In the learning process, soft skills should add to the content of the STEM paradigm for students to stay competitive in Industry 5.0 in the future.

5 Design with STEM Education

STEM education is an interdisciplinary field of study integrating science, technology, engineering, and mathematics. More than 14 countries have involved STEM in their curriculum since the concept was first introduced by the United States. After Obama's speech on STEM Education, there were seen efforts in the US [11], and EU has also made a similar effort in this area. They focus more on STEM project creation, like Scientix and STEM Alliance. Besides these, many scholars contribute various pedagogical methods or frameworks to improve the efficiency and quality of STEM education. In this study, we proposed a combination of STEM and Design to explore a new form of teaching and learning.

Design can be considered as a completed plan to construct an object or system in a creative, usable and efficient process. The definition of Design is a decision upon the figure and functioning of a product, building, or other objects by making a detailed description of it. [12]. There are myriad philosophies that guide design as design values and related aspects of modern design fields between different institutions of thought and practicing designers. In this way, design is at the heart of all vocational training. It is the main symbol that distinguishes professions from science and technology. As a field of research, design comprehends not only design disciplines but also manufacturing systems, economic frameworks, information technologies, and educational platforms. Design theory and skills as the basis of in next industrial generation.

Nowadays, design has undergone a series of evolutions over five industrial revolutions. At the beginning of the 20th Century, the Bauhaus school has highly influenced almost all design schools in the world in Germany. Walter Gropius, who is a famous architect, became the founder of the Bauhaus school. The school integrated the crafts and fine arts to propose a specific approach intended to unify individual artistic vision with the principles of flow production and an emphasis on everyday functionality. The students in this design school accepted artistic learning and crafts manufactural training. A breakthrough of Gropius, modern art training in industrial society should be changed into a scientific and creative method to improve both the technical knowledge and soft skills of students. Usability and functionality are the two most important conditions in product design [13]. The traditional art model tends to have the individual artist complete the craft pieces. The educators in this school separated the art from the isolated mode to collaborate the skills of craftsmen and artists in the project [14]. The school proposed to elevate the status of artware to the equivalent

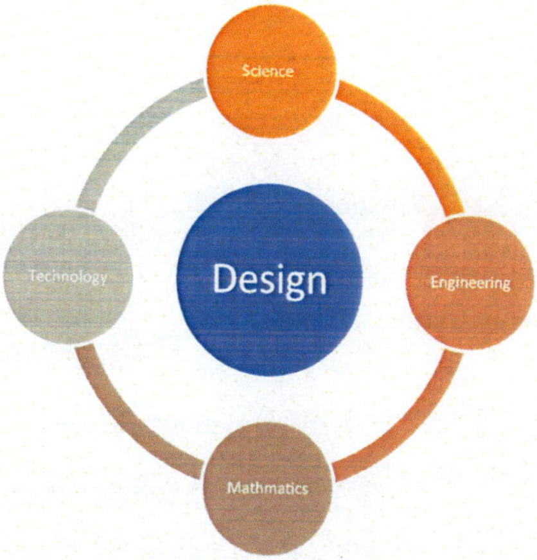

Fig. 1 Basic framework of D-STEM

level as itself, such as painting and sculpture. In addition, the school began by maintaining communication with the leaders of industry and crafts to export innovative masterpieces to industry and create value for the government. Thenceforth, the relationship between design and industry is getting closely combined. In entering the fifth industrial revolution, design thinking and ability can be used as a meaningful medium named Design-STEM (D-STEM) education (Fig. 1).

Furthermore, the research related to design fields, and the researcher chose STEM to be the object of study. Thus, it is necessary to understand the engineering models of product design. One of the most essential models in the engineering design process is the systematic approach by Pahl and Betiz form 1988 in which the flow of work during the design process is divided into four main steps: Clarification of the task, Conceptual design, Embodiment design, and Detail design [15]. Therefore, a specific design process model (Fig. 2) focused on STEM has been generated based on Pahl and Betiz's model. This specific design process model focused on the "Design-STEM" in education. The model created for the design integration of science and technology, and the process should conform to the current situation of the industry development. The design process model involves four main steps. Definition of the task: this is the first step of a design project. Students need to understand the aim of the task and narrow down the problem. Find out the scientific concept and technology factor. It will build their ability to identify and solve problems. Conceptual Design: it is a framework in an early phase of the design process that students to figure out a solid solution. A prototype always is created in this step to export the design idea and target of the project. Design Representation: a simulation model generated in a 3D modelling software; the 3D printer is used to export the model in fast speed. Design Enhancement: evaluate the product features and solve defects before the product is mass-produced [16].

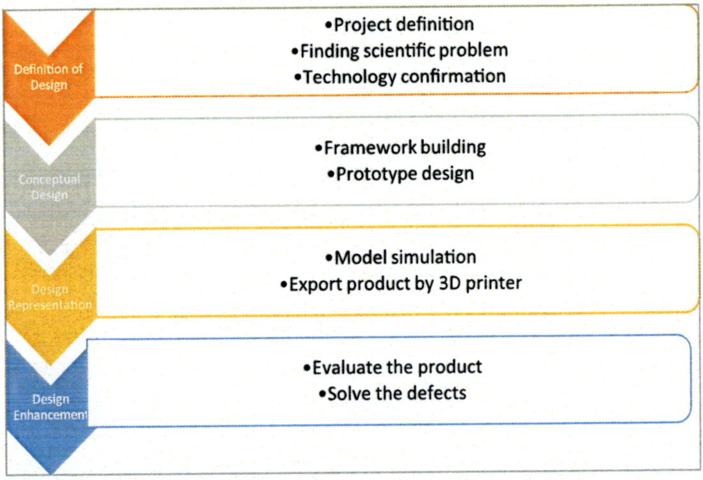

Fig. 2 Design process model for STEM education

6 Design Technology with 3D Optical Scanning and Modeling

The combination with advances technology and design method is a widespread issue for scholars and the whole science and technology fields. The appearance of 3D optical technology provides a new approach to Design-STEM education.

3D scanner used in this research is a new technology that combines optical, mechanical, electrical and computer technology which is mainly used to scan the spatial shape and structure of an object to obtain each coordinate. This technology plays a fundamental role as an important technical support in the development of manufacturing, and is gradually developing into a new discipline in the application [16]. The significance technology can convert the three-dimensional information of the object into a digital signal that can be directly processed by a computer, which provides a convenient method for digitizing the object. The use of 3D and AI technology in education is not a new method; we focus on integrating 3D optical technology and D-STEM to show the potential and application prospects in this multi-disciplinary subject. This technology with fast speed and a wide range of high-accuracy data brings beneficial influence and innovation to design, science, technology, education, and industry.

3D scanning technology has developed several scanning principles, which are basically divided into two main categories: contact (probe) and non-contact (laser, photographic, X-ray) since its inception. The features of the contact scanner are low cost but slower speed. There will be blind spots when the probe contacts the object in a soft texture environment. Contact 3D scanners have high accuracy for relative size measurement and quality management. Optical scanning through fast speed and appropriate accuracy to obtain the amount of point clouds data to support surface reconstruction.

Many companies focus on the development of 3D scanning technology to integrate two or more technologies to form a unique composite 3D scanning system, which can measure three-dimensional data and the two-dimensional contour of the objects simultaneously. This approach has great potential for 3D scanning technology, which combines the strengths of various technologies and aims to avoid the imperfection of a single measurement method. A widespread application of high-speed 3D scanning systems in reverse engineering, industrial manufacturing, clothing production, facial surgery, terrain investigation, urban planning, and digital entertainment fields. They have been applied in automobiles, motorcycles, and home accessories successfully to change the traditional industry into an innovative mode.

The core technology of the 3D scanner in our research consists of a multiple stereo vision camera and structured light illuminating equipment, which is based on the fundamental structure of a 3D sensor with phase-mapped structured light illumination. The equipment used to capture an object's in-depth 3D modelling and colour texture data in several seconds long to obtain complete data from automated stitching of three-dimensional data. The complete data can be exported to modelling software such as AutoCAD and Inventor for the next creation step. Because 3D optical technology can measure and record the data of complex objects, students are able to re-design the different colours and surfaces based on existing models captured by the 3D scanner (Fig. 3).

3D optical technical hardware with the design process to improve cognitive and creative ability during STEM learning. By advancing the science of design and by creating a broad computer-based methodology for automating the design of artefacts and industrial processes, scholars can attain dramatic productivity improvements [17]. The 3D optical technology introduced in this study allows educators and students to perform how this technology impacts cognition and creativity in learning and teaching.

Fig. 3 3D optical scanner and model

Fig. 4 Example of innovation laboratories

Many schools in China have started to build innovation laboratories to prepare for Industry 5.0. Shenzhen Mingde Experimental Schools raised the philosophy of "Opening the boundary, Integrating the future," which means opening the campus boundary to integrate the government and enterprises. Eight composite school-enterprise joint innovation laboratories in 2021 in cooperation with enterprises such as Dji, Tencent, UBTECH Robotics, BYD, BGI Group, CECEP Tech and Ecology, CIMC, Han's Laser Technology, and Shenzhen Water and Environment Group. The laboratories include Aerospace, Automotive Engineering, Bionic Machinery, Digital Manufacturing, Sponge City Lab, Biogenetic, and Marine Resources, which relate to Intelligence Manufacturing, Ecological Nature, Medica Art, Science of Material, and Quantum Technology (Fig. 4). This program has been used by more than 4000 students of all levels who have participated in competing eight thematic technology exploration courses.

7 Conclusion

This study aim to use 3D optical with Artificial Intelligence technology to assist STEM teaching and learning. The advanced technology is able to improve the practicing and modelling ability for students when they just start science and technology study. We insert "Design" in the study process that has characteristics in common. Both of them connect with interdisciplinary, including cognitive, collaborative, and pedagogical representations of students' behavior and creative skill that enable a better understanding of students. The D-STEM education is more in line with the current situation of Industry 4.0 and Industry 5.0 in the future. We require students not to lose their creativity skills. For industrial talent to be educated in the twenty-first century, they need to study both design integrated science and technology through their schooling [18]. Even now and future, people need both design and scientific thinking to balance them with the criticality interactively, cognitively, and creativity of the mind when working with smart machines in the advanced Industry 5.0 era.

References

1. Wells, J.G.: STEM education: the potential of technology education. In: Paper presented at The Mississippi Valley Conference in the 21st Century: Fifteen Years of Influence on Thought and Practice
2. Cropley, A.: Creativity-focused technology education in the age of industry 4.0. Creativity Res. J. **32**(2), 184–191 (2020)
3. Kärcher, B.: Alternative Wege in die Industrie 4.0–Möglichkeiten und Grenzen. In: Zukunft der Arbeit in Industrie 4.0, pp. 47–58. Springer Vieweg, Berlin, Heidelberg (2015)
4. Javaid, M., Haleem, A., Singh, R.P., Suman, R.: Artificial intelligence applications for Industry 4.0: A literature-based study. J. Ind. Integr. Manage. **07**(01), 83–111 (2022). https://doi.org/10.1142/s2424862221300040
5. Ribeiro, J., Lima, R., Eckhardt, T., Paiva, S.: Robotic process automation and artificial intelligence in Industry 4.0–A literature review. Proc. Comput. Sci. **181**, 51–58 (2021)
6. Mazzone, M., Elgammal, A.: Art, creativity, and the potential of artificial intelligence. Arts **8**(1), 26 (2019). https://doi.org/10.3390/arts8010026
7. Maddikunta, P.K.R., Pham, Q.-V., B, P., Deepa, N., Dev, K., Gadekallu, T.R., Ruby, R., Liyanage, M.: Industry 5.0: A survey on enabling technologies and potential applications. J. Ind. Inf. Integr. **26**, 100257 (2022). https://doi.org/10.1016/j.jii.2021.100257
8. Commission, E.: Enabling technologies for Industry 5.0-results of a workshop with Europe's technology leaders (2020)
9. Gürdür Broo, D., Kaynak, O., Sait, S.M.: Rethinking engineering education at the age of industry 5.0. J. Ind. Inf. Integr. **25** (2022). https://doi.org/10.1016/j.jii.2021.100311
10. Larson, L.C., Miller, T.N.: 21st century skills: prepare students for the future. Kappa Delta Pi Rec. **47**(3), 121–123 (2011). https://doi.org/10.1080/00228958.2011.10516575
11. Idin, S.: An overview of STEM education and Industry 4.0. In: Mack Shelley, S.A.K. (ed.) Research highlights in STEM education. pp. 194–208. International Society for Research in Education and Science, Iowa State University, 509 Ross Hall, Ames, IA 50011-1204, U.S.A. (2018)
12. Macdonald, S.: The history and philosophy of art education. James Clarke & Co. (2004)
13. Jiang, H., Tang, M., Peng, X., Liu, X.: Learning design and technology through social networks for high school students in China. Int. J. Technol. Des. Educ. **28**(1), 189–206 (2018). https://doi.org/10.1007/s10798-016-9386-8
14. Whitford, F.: Bauhaus Thames and Hudson. London, UK, pp 29e30 (1984)
15. Pahl, G., Beitz, W.: Engineering design: a systematic approach. NASA STI/RECON Techn. Rep. A **89**, 47350 (1988)
16. Jiang, H., Liu, X.-L., Peng, X., Tang, M.-X.: An interactive model of creative design behavior with 3D optical technology. In: Design, user experience, and usability: design thinking and methods, pp. 43–52. Springer International Publishing (2016)
17. Amarel, S.: Artificial intelligence and design. In: Proceedings of the 5th Jerusalem Conference on Information Technology, 1990. 'Next Decade in Information Technology', pp. 315–333. IEEE (1990)
18. Braund, M., Reiss, M.J.: The 'Great Divide': How the arts contribute to science and science education. Can. J. Sci. Math. Technol. Educ. **19**(3), 219–236 (2019). https://doi.org/10.1007/s42330-019-00057-7

Enhancing Engineer and Engineering Perception Through Video Design in STEM Education

N. Muñoz-Rujas, A. Pavani, J. Baptiste, F. E. M. Alaoui, and E. Montero

Abstract Engineering is an activity of great importance for societies as it facilitates problem solving and innovation. Very often, however, societies have a limited understanding of what engineers do. Frequently, engineering is seen only as application of science and the use of technological tools. But engineering is more than this, it involves creativity (problem scoping and identifying multiple solutions), evaluation (selecting, testing, and improving solutions) and materializing optimal human-made solutions in real life. Several studies on engineer and engineering perceptions of teachers and students in STEM education (Science, Technology, Engineering, and Mathematics) reveal narrow and stereotypical conceptions that lead to a misconception of Engineering. We present here an initiative which aim to broaden the knowledge of engineers as lead actors of many well-known social improvements. The project is addressed to teachers of primary education, secondary education (K-12), and engineering lecturers, providing them with a set of case studies that can be used for classroom assignments or open discussion within the frame of any STEM education program. Preliminary results presented in this contribution consist of a set of videos connecting engineers and day-to-day appliances and include three development steps: context of the case study for open discussion, post of the main question

N. Muñoz-Rujas (✉) · E. Montero
Universidad de Burgos, Burgos, Spain
e-mail: nmrujas@ubu.es

E. Montero
e-mail: emontero@ubu.es

A. Pavani
Pontifícia Universidade Católica do Rio de Janeiro, Rio de Janeiro, Brazil
e-mail: apavani@puc-rio.br

J. Baptiste
University of Trinidad and Tobago, Port of Spain, Trinidad and Tobago

F. E. M. Alaoui
Université Abdelmalek Esaadi, Tetouan, Morocco
e-mail: f.mhamdialaoui@uae.ac.ma

L. F. M. da Silva and A. J. M. Ferreira (eds.), *3rd International Conference on Science and Technology Education 2022*, Proceedings in Engineering Mechanics,
https://doi.org/10.1007/978-3-031-25401-7_3

to solve, and how engineers engineered the solution within the economic and cultural context of their time.

Keywords Engineers · STEM education · Video design · K-12 · Problem solving

1 Perception of Engineering in STEM Education

According to the traditional definition, engineering has been understood as an application of mathematics and science. This is a very narrow point of view which is far from reality. More recent definitions, as the one given by the American Society for Engineering Education [1], consider that "engineers solve problems using science and math, harnessing the forces and materials in nature. They draw on their creative powers to come up with quicker, better, and less expensive ways to do the things that need to be done". The main difference between both definitions is the idea that solving problems comes first, the engineer aims to solve problems concerning society.

Engineering is nowadays conceptualized in education by its connection to other disciplines. In recent times, the graduate education system for science, technology, engineering, and mathematics (STEM) has received increased attention from society and governments. In the last decades [2, 3] STEM is still being reformulated from experience [4, 5]. Most studies in the literature refer to general reviews of the state of progress of STEM initiatives, report STEM implementation in certain engineering or science courses, or are case studies in limited contexts, by country or educational level. Nevertheless, some authors have pointed out some misconceptions about engineering in STEM activities.

Many STEM proposals tend to centre the focal point on the use of technology per se, instead of highlighting the relevance of the interaction in the social context. Thus, engineering, technology education, and STEM education, have often been criticized for presenting a techno-centric view. For instance, seeing the use of software as identified mainly as computing as the main concern, producing a distortion of the comprehension of the technology in STEM [6].

Being aware that there are epistemological differences between engineering, technology, and science [7], some experts propose that engineering education should follow the example of science education, that is: paying more attention to the process than to the products [8]. Emphasizing engineering as a social and cultural activity [9], means that it covers a specific behaviour while doing and thinking [10]. The interesting work of Pleasants and Olson [11] presents the key dimensions of what engineering is and what engineers do through a compilation of historical, philosophical, and sociological perspectives, as well as from an engineering perspective.

As far as the understanding of engineering is concerned, society generally has a limited and confusing knowledge of what engineering is and what engineers actually do. Several studies indicate that teachers' perceptions and attitudes towards engineering can affect those of their students [12]. The work of Kuvac and Koc [13] points out that many teachers possess a stereotypical idea of engineering as a male profession that involves working alone. Some teachers tend to marry the idea of the engineer with the physical task of building a machine or a vehicle. Another study declares that many teachers have little knowledge about engineering and engineers, and that they are not able to distinguish between examples of science activities and engineering activities [14]. Some of these works investigate the effectiveness of some interventions on teachers to make their perceptions of engineers and engineering less biased and more accurate [13].

In parallel, there is also a diversity of studies on the perception that students have of engineers and engineering. The work of Capobianco et al. [15] reveals that elementary school students have a perception of engineering described primarily as fixing, building, and working on vehicles, engines, and tools. When they draw an engineer, they describe him as a manual worker, usually a male. Another study makes a comparison of what similarities and differences middle school students perceive between a scientist and an engineer [16]. Using a methodology of drawing a scientist and drawing an engineer, the work shows that a high fraction of students has no perception of what an engineer is. Those who do manifest some knowledge describe him rather as a manual worker in an outdoor space, while the scientist is described by conducting experiments in an indoor space (laboratory), being in both cases of male gender.

Following the same method of drawing an engineer along with individual interviews, another study reveals that the concepts middle school students have about engineering and engineers is uncertain. Some students noted that engineers design and create products and others focused more on making or building the product. In any case, the dominant perception is that engineering is something done by a skilled craftsman [17].

If the aim is the true integration of STEM education so that its practical application is effective and balanced, there is a need to carry out more studies, propose training actions and promote materials to improve the perception of engineering and engineers among teachers and students. This paper presents the initial results of a project whose objective is to make versatile and freely available materials for use by teachers who want to use real-life examples to introduce the basic concepts of engineering in STEM contexts.

2 Promotion of Engineering Perception Through Video in STEM Education

The use of video as an educational tool has increased enormously in recent years in all disciplines: literature, art, history, science, technology, etc. Video can be used not only for teaching, but also for studying and learning a particular topic in and outside the classroom. It is important not to fall into the trap of considering video as a magic solution that will make students learn more or be more motivated. Simply presenting information in an interesting video does not automatically mean that it will lead to more meaningful and deeper learning [18]. Videos are only one component in the whole activity of the education system. Learning outcomes depend largely on how videos are used as part of the overall approach to learning. Watching videos must be integrated into the rest of resources and tasks to be performed.

The design of a video for teaching should be done according to well-founded pedagogical choices. It is up to its authors to decide in advance their components and characteristics. There are studies that define a framework for categorizing the uses of videos according to learning outcomes [19]:

- Seeing: A video can help people see things they may not have seen before and help them by pointing out details that these people wouldn't have noticed otherwise.
- Engaging: A video develops the student's interest by participating in a learning activity. The development of intrinsic motivation can be done by biting into the student's curiosity or showing relevance to the real world.
- Doing: This implies both a change of attitude and the acquisition of skills.
- Saying: This requires the student to acquire verbal or declarative knowledge, e.g., facts, explanations: people remember facts better when these facts are the solution to a problem that a person has reached, rather than giving an isolated statement.

Therefore, more than the medium, the relevant aspects to design a video are the pedagogy, a well-structured message, the overall approach and the design. Thc teacher's task is to offer a coherent path through the video so that the student is aware of the learning in the activity in which he is participating.

3 Case Study and Discussion

The case presented below is part of the project Engineers and STEM [20]. The Project has the support of MERLOT (Multimedia Education Resource for Learning and Online Teaching) [21]. MERLOT is one of the most reputed Open Educational Resources repository devoted to identifying, peer reviewing, organizing and making available existing online learning resources in a wide range of academic disciplines.

The objective of the project Engineers and STEM is to contribute to STEM promotion worldwide by highlighting the role of engineer's contribution to the development of well-known social, scientific, and technological developments. We expect to improve the perception of engineers and engineering amongst teachers and students. The results of the project will consist of a set of exercises, discussion topics, reference documents, videos and questionnaires on Engineering and Society for teachers of STEM subjects. These materials will be allocated in a public website, still in development. The project is aimed at teachers of primary education, secondary education (K-12), and engineering lecturers. The website materials could be used for free by teachers at any level, as they could decide how to integrate them in their respective teaching approach.

At present, the website offers three case-studies, and we will present in this contribution the design criteria of the respective three videos involved. The website contains the context, the example, the suggestions for open discussion, the list of external links to broaden information and the link to a self-assessment questionnaire accompanying the respective video. In this way we accomplish the design criteria described in the previous section, concerning the video and the meaningful teaching context. The selection criteria of the examples to be included in the project is that they must be well-known social, scientific, and technological developments which have a direct relationship with an engineer.

Even though the videos are embedded on the website, they are also available in YouTube as stand-alone materials. As the project has an international scope, we have developed every video in several languages: English, Spanish, Portuguese, and French. Table 1 presents the list of case-studies.

Before we began to develop the videos, we did take into consideration the following basic questions:

- Who is our audience?
  Our audience are those teachers that teach any STEM discipline at primary or secondary education (K-12), and at engineering level, who want to integrate a better understanding of the E of STEM. They could find any case study from the repertory posted on the website that fits their educational program.
- What is the learning objective and goal for using the video as an alternative to other presentation formats?
  The video could be very useful as a compact presentation of the engineer behind a well-known fact or product, to highlight the impact of engineers on society. It seems video presentations are more effective than written portrayals for presenting

Table 1 List of 3 case-study videos in MP4 format of the project Engineers and STEM

Language	EN	ES	PT	FR
Case	1. Who defined the *horsepower* as the unit of power?			
URL	https://youtu.be/2sUGNJbiCS0	https://youtu.be/oIiIjdmt7KM	https://youtu.be/v5Ra23KahwQ	https://youtu.be/OPEDl2rwYsQ
Duration min:sec	3:57	3:59	3:58	3:45
Case	2. Did James Watt define the *watt* unit for power with his own name?			
URL	https://youtu.be/shDppTco3KE	https://youtu.be/AHy-WiopIlg	https://youtu.be/DaK4NRFuGfM	https://youtu.be/6e0mvlRLRxk
Duration min:sec	4:31	4:26	4.24	4.13
Case	3. Some temperatures are measured in degrees Rankine. Who was Mr. Rankine?			
URL	https://youtu.be/Qj9CwqdlnUQ	https://youtu.be/YZn2HKftXOs	https://youtu.be/5eG2m0-kj68	https://youtu.be/HkWYNRLz124
Duration min:sec	4:42	4.24	4.36	4:36

scientists or engineers [22]. To keep the attention of the viewer, it must be of seven (7) minutes duration or less [23]. Reading texts are included in some sections, as verbal redundancy of listening to reinforce the main ideas [24]. The video could be reused and shared several times.

- How will we/the teachers know if we have successfully conveyed the ideas?
  The teacher can propose open discussion sessions or writing reports on the topic before and after watching the video to evaluate the impact on the students.
  Included on the website, are self-assessment questionnaires to evaluate student understanding.
- When the video will be presented to students?
  It is the teacher who will decide when and how to use the available materials integrating them into his or her course map of the education program. (As pre-work before class? In-class, as a way to gain the attention of the class or as a case study for discussion? As a study resource after the class?).
- What materials do we already have prepared?
  Using appropriate and appealing pictures and visuals is of utmost importance to maintain viewer interest. First, we collect available information about the case study and then we select the most effective for the video. Most of our case-studies refer to well-known facts that come from historical engineers, hence, original documents and the background of the engineers involved have been used. We avoided using trademarks or commercial names linked to the respective engineer because of the copyright concern. We prefer to use Public Domain documents and pictures, under Creative Commons license [25].
- What resources do we have to create the video?
  We have used licensed software (academic or professional) to edit the videos, recording the audios, preparing presentations, and editing the pictures. Once the videos are ready, they are uploaded to the open repository of educational resources (RIUBU) and to the YouTube channel of the University of Burgos, Spain. [26]
- How will the teachers incentivize students to actually watch, process, and reflect on the information in the video?
  On the website associated to the video, there are open discussion issues for students, and links to the Internet to broaden information.

According to these design principles and the cited basis of Schwartz and Hartman [19], the videos were edited with the following structure:

1. Declaration of a well-known fact or product directly related to an engineer. Presentation of the evidence through an example to draw people's attention to a detail that usually goes unnoticed.

Case 1: The horsepower, hp, as unit of power.

When we check a brochure of a car, one of the most relevant characteristics, which influences the decision to acquire the vehicle, is the power of the engine. We see in the example shown, that the power of the engine is expressed in "horsepower", with the symbol hp (Fig. 1).

	FWD, ECOTEC 1.3L TURBO ENGINE
Powertrain	
Drivetrain	Front Wheel Drive
Standard Engine	Gas I3
Fuel Injection	Direct Injection
Horsepower @ RPM	155 @ 5600
Torque @ RPM	174 @ 1600
Transmission, Continuously Variable (CVT)	Standard
Transmission, 9-speed automatic	

Fig. 1 Technical sheet of a car with power expressed in horsepower

Case 2: The watt, unit of power of the International System of Units.

We are used to having many household appliances, which help us to make our life easier. Let's turn around these devices and we will find a technical label. In all of them we can read a line with the symbol W, which we name "watt". It tells us something about the power of the device. As he is very famous worldwide, many people could associate this name to the Scottish engineer James Watt (Fig. 2).

Fig. 2 Picture of a toaster and technical label with mention of the power in W

Case 3: The Absolute Temperature Rankine scale.

In some US engineering textbooks dealing with temperature concepts, we find the mention of a temperature measured in a so-called Rankine scale. Very often, the relations between this temperature scale and those of Celsius, Fahrenheit and Kelvin are described by the respective conversion factors (Fig. 3).

2. Asking for the engineer behind this well-known fact or product.

Case 1: Who defined the horsepower as the unit of power?
Case 2: Did James Watt define the watt unit for power with his own name?
Case 3: Who was that Mr. Rankine of the Rankine temperature scale?

3. Presentation of the engineer in short, an explanation of the link between the well-known fact or product and the engineer, how the engineer reached the solution to a practical problem or how his or her contribution was recognized by others by naming this fact or product with the engineers' name.

2.5.4 Celsius, Rankine, and Fahrenheit Scales

triple point

Temperature scales are defined by the numerical value assigned to a *standard fixed point.* By international agreement the standard fixed point is the easily reproducible ***triple point of water:*** the state of equilibrium between steam, ice, and liquid water (Sec. 4.2). As a matter of convenience, the temperature at this standard fixed point is defined as 273.16 kelvins, abbreviated as 273.16 K. This makes the temperature interval from the *ice point*[1] (273.15 K) to the *steam point*[2] equal to 100 K and thus in agreement over the interval with the Celsius scale that assigns 100 Celsius degrees to it.

Celsius scale

The ***Celsius temperature scale*** (formerly called the centigrade scale) uses the unit degree Celsius (°C), which has the same magnitude as the kelvin. Thus, temperature *differences* are identical on both scales. However, the zero point on the Celsius scale is shifted to 273.15 K, as shown by the following relationship between the Celsius temperature and the Kelvin temperature:

$$T(°\mathrm{C}) = T(\mathrm{K}) - 273.15 \tag{2.14}$$

From this it can be seen that on the Celsius scale the triple point of water is 0.01°C and that 0 K corresponds to −273.15°C.

Rankine scale

Two other temperature scales are in common use in engineering in the United States. By definition, the ***Rankine scale,*** the unit of which is the degree rankine (°R), is proportional to the Kelvin temperature according to

$$T(°\mathrm{R}) = 1.8T(\mathrm{K}) \tag{2.15}$$

As evidenced by Eq. 2.15, the Rankine scale is also an absolute thermodynamic scale with an absolute zero that coincides with the absolute zero of the Kelvin scale. In thermodynamic relationships, temperature is always in terms of the Kelvin or Rankine scale unless specifically stated otherwise.

Fahrenheit scale

A degree of the same size as that on the Rankine scale is used in the ***Fahrenheit scale,*** but the zero point is shifted according to the relation

$$T(°\mathrm{F}) = T(°\mathrm{R}) - 459.67 \tag{2.16}$$

Substituting Eqs. 2.14 and 2.15 into Eq. 2.16, it follows that

$$T(°\mathrm{F}) = 1.8T(°\mathrm{C}) + 32 \tag{2.17}$$

This equation shows that the Fahrenheit temperature of the ice point (0°C) is 32°F and of the steam point (100°C) is 212°F. The 100 Celsius or Kelvin degrees between the ice point and steam point correspond to 180 Fahrenheit or Rankine degrees.

Fig. 3 Section of an engineering textbook dealing with temperature scales

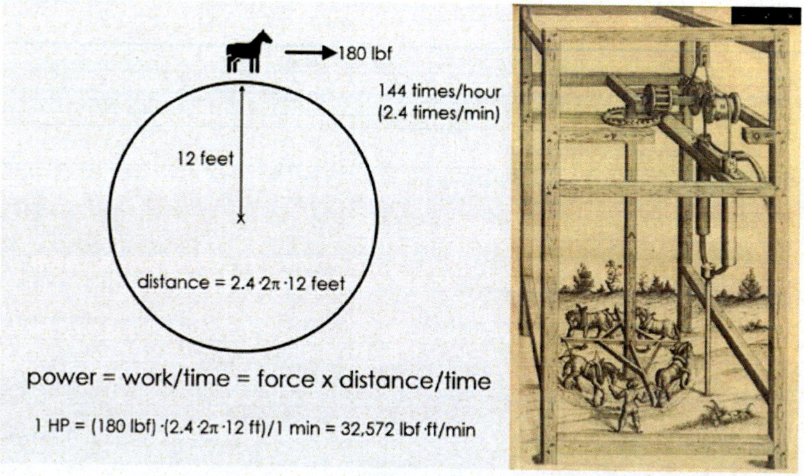

Fig. 4 Definition of the horsepower by Watt: how much power can a standard horse develop

Case 1: The horsepower as unit of power.

The "horsepower" was named by the engineer James Watt in 1809, who defined the steam engine and its qualities in a commercial and clever form. Since the steam engine had to compete, at that time, with the horse like source of energy in the mining industry, he decided to use the horse as a unit of measure. Was there a better form to present the steam engine qualities than indicating the number of horses that it could substitute? That is the origin of the name "horsepower" (Fig. 4).

Case 2: The watt, unit of power of the International System of Units.

Watt didn't define the watt, W, which is the unit of power of the International System of Units. The watt as a unit of power was proposed and defined in 1882 by the German electrical engineer Carl Wilhelm Siemens and was initially related to the electrical properties because of the strong development of electricity as source of power: 1 W = 1 A × 1 V. The unit was given the name of watt by the British Association for the Advancement of Science in honour of the engineer James Watt (Fig. 5).

Case 3: The Absolute Temperature Rankine scale.

The Scottish mechanical and civil engineer, and professor at the University of Glasgow, William Rankine, published in 1859 his Manual of the Steam Engine and other Prime Movers. Rankine introduces the definition of the Absolute Temperature scales in relation to the ordinary scales, Celsius and Fahrenheit. In relation with the Fahrenheit scale, he proposes the respective Absolute Temperature where the absolute zero corresponds to −461.2°F (Today value is −459.67°F). This absolute temperature scale was named afterwards as the Rankine Scale (Fig. 6).

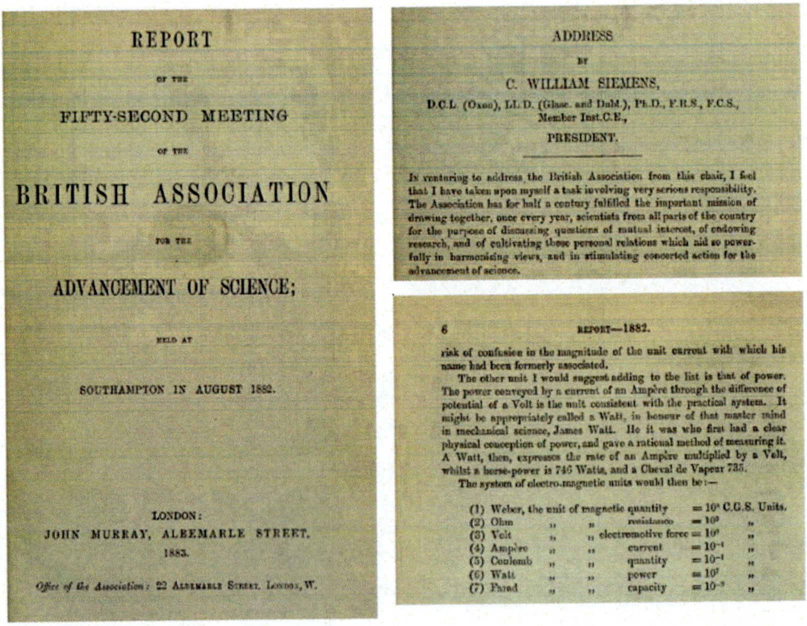

REPORT

OF THE

FIFTY-SECOND MEETING

OF THE

BRITISH ASSOCIATION

FOR THE

ADVANCEMENT OF SCIENCE;

HELD AT

SOUTHAMPTON IN AUGUST 1882.

LONDON:
JOHN MURRAY, ALBEMARLE STREET.
1883.

Office of the Association: 22 ALBEMARLE STREET, LONDON, W.

ADDRESS

BY

C. WILLIAM SIEMENS,

D.C.L. (Oxon), LL.D. (Glasg. and Dubl.), Ph.D., F.R.S., F.C.S.,
Member Inst.C.E.,

PRESIDENT.

IN venturing to address the British Association from this chair, I feel that I have taken upon myself a task involving very serious responsibility. The Association has for half a century fulfilled the important mission of drawing together, once every year, scientists from all parts of the country for the purpose of discussing questions of mutual interest, of endowing research, and of cultivating those personal relations which aid so powerfully in harmonising views, and in stimulating concerted action for the advancement of science.

6 REPORT—1882.

risk of confusion in the magnitude of the unit current with which his name had been formerly associated.

The other unit I would suggest adding to the list is that of power. The power conveyed by a current of an Ampère through the difference of potential of a Volt is the unit consistent with the practical system. It might be appropriately called a Watt, in honour of that master mind in mechanical science, James Watt. He it was who first had a clear physical conception of power, and gave a rational method of measuring it. A Watt, then, expresses the rate of an Ampère multiplied by a Volt, whilst a horse-power is 746 Watts, and a Cheval de Vapeur 735.

The system of electro-magnetic units would then be:—

(1) Weber,	the unit of	magnetic	quantity	$= 10^{8}$	C.G.S. Units.
(2) Ohm	"	"	resistance	$= 10^{9}$	"
(3) Volt	"	"	electromotive force	$= 10^{8}$	"
(4) Ampère	"	"	current	$= 10^{-1}$	"
(5) Coulomb	"	"	quantity	$= 10^{-1}$	"
(6) Watt	"	"	power	$= 10^{7}$	"
(7) Farad	"	"	capacity	$= 10^{-9}$	"

Fig. 5 Report presented in 1882 by Carl Wilhelm Siemens, president of the British Association for the Advancement of Science, proposing and naming the unit of power as watt in honour of the engineer James Watt [27]

4. As a closure, the presentation of some additional information, usually unknown, about the engineer serves to bite the curiosity of the viewer: impact on the society, and on the science and technology system; additional vision of the engineer profile; showing how science, engineering and technology interact.

Case 1: The horsepower as unit of power.

The "horsepower" was deeply meaningful for his fellow citizens. Watt not only improved the steam engine as a technical alternative for the mining industry but comes to easy the understanding for the general people. The "horsepower" was born close to people experience with horses, the main source of power at that time. It will be one of the latest units of measurement coming from the traditional man experience (foot, inch…) before the metric system came into the scenery (Fig. 7).

Case 2: The watt, unit of power of the International System of Units.

There are many scientific units named after people. Much of the personal naming occurred in 1860–70s for CGS system and 1870–1880s for the MKS system. The most frequent case is that the unit is given its name by someone else to honour the discoverer or because nobody could come up with a better name. By convention, the

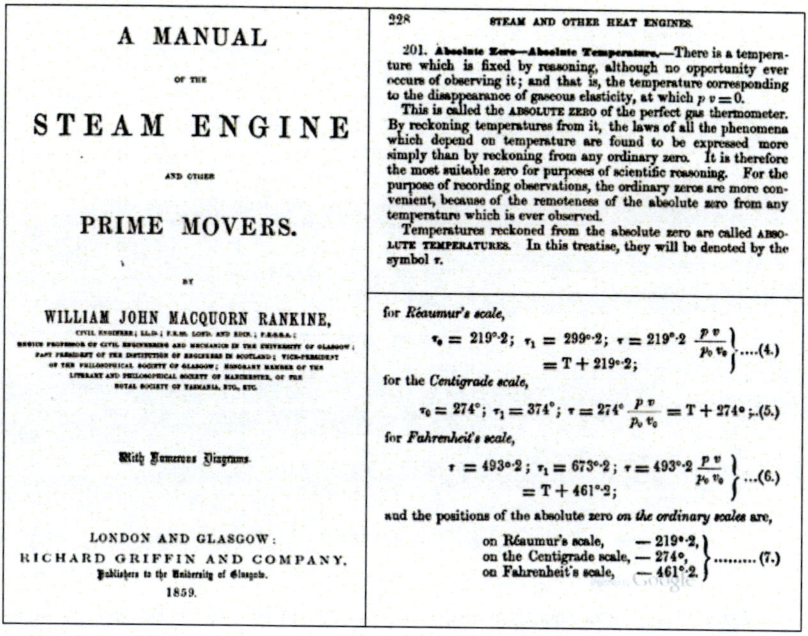

A MANUAL

OF THE

STEAM ENGINE

AND OTHER

PRIME MOVERS.

BY

WILLIAM JOHN MACQUORN RANKINE,

With Numerous Diagrams.

LONDON AND GLASGOW:
RICHARD GRIFFIN AND COMPANY,
Publishers to the University of Glasgow.
1859.

228 STEAM AND OTHER HEAT ENGINES.

201. **Absolute Zero—Absolute Temperature.**—There is a temperature which is fixed by reasoning, although no opportunity ever occurs of observing it; and that is, the temperature corresponding to the disappearance of gaseous elasticity, at which $p\,v = 0$.

This is called the ABSOLUTE ZERO of the perfect gas thermometer. By reckoning temperatures from it, the laws of all the phenomena which depend on temperature are found to be expressed more simply than by reckoning from any ordinary zero. It is therefore the most suitable zero for purposes of scientific reasoning. For the purpose of recording observations, the ordinary zeros are more convenient, because of the remoteness of the absolute zero from any temperature which is ever observed.

Temperatures reckoned from the absolute zero are called ABSOLUTE TEMPERATURES. In this treatise, they will be denoted by the symbol τ.

for *Réaumur's scale*,

$$\left.\begin{array}{l}\tau_0 = 219^\circ{\cdot}2;\ \tau_1 = 299^\circ{\cdot}2;\ \tau = 219^\circ{\cdot}2\,\frac{p\,v}{p_0\,v_0}\\ = \mathrm{T} + 219^\circ{\cdot}2;\end{array}\right\}\ldots(4.)$$

for the *Centigrade scale*,

$$\tau_0 = 274^\circ;\ \tau_1 = 374^\circ;\ \tau = 274^\circ\,\frac{p\,v}{p_0\,v_0} = \mathrm{T} + 274^\circ;\ldots(5.)$$

for *Fahrenheit's scale*,

$$\left.\begin{array}{l}\tau = 493^\circ{\cdot}2;\ \tau_1 = 673^\circ{\cdot}2;\ \tau = 493^\circ{\cdot}2\,\frac{p\,v}{p_0\,v_0}\\ = \mathrm{T} + 461^\circ{\cdot}2;\end{array}\right\}\ldots(6.)$$

and the positions of the absolute zero *on the ordinary scales* are,

$$\left.\begin{array}{ll}\text{on Réaumur's scale,} & -219^\circ{\cdot}2,\\ \text{on the Centigrade scale,} & -274^\circ,\\ \text{on Fahrenheit's scale,} & -461^\circ{\cdot}2.\end{array}\right\}\ldots\ldots\ldots(7.)$$

Fig. 6 Section of the engineering textbook of W. Rankine with the calculations of absolute temperatures [28]

Fig. 7 Comparison between the steam engine and a horse turning a wheel: everybody could understand this picture

name of the unit in honour of a person is properly written in all-lowercase, but its abbreviation is capitalized (Fig. 8).

Case 3: The Absolute Temperature Rankine scale.

Though scarcely used, the Rankine scale is still in use in the US within the aerospace industry because there are a lot of programs that were developed during the second half of the twentieth century using Rankine. Thus, it seems that, to be compatible with those old programs, it's often simpler to just use Rankine in the new programs too (Fig. 9).

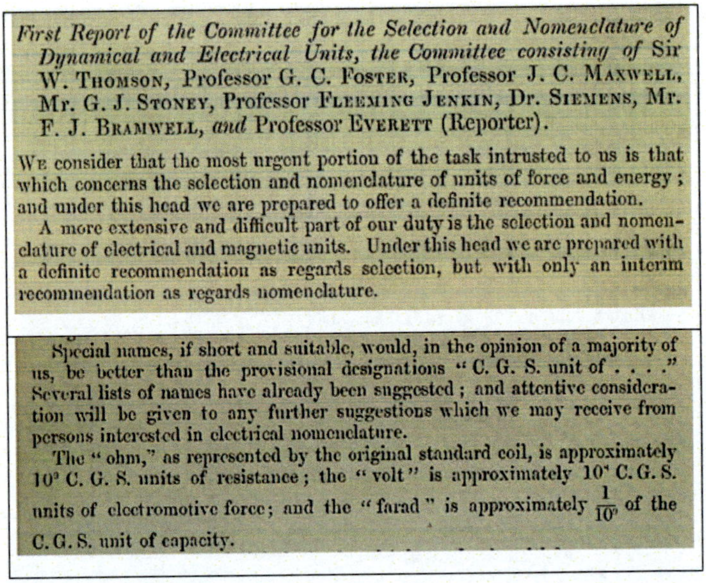

First Report of the Committee for the Selection and Nomenclature of Dynamical and Electrical Units, the Committee consisting of Sir W. THOMSON, Professor G. C. FOSTER, Professor J. C. MAXWELL, Mr. G. J. STONEY, Professor FLEEMING JENKIN, Dr. SIEMENS, Mr. F. J. BRAMWELL, *and* Professor EVERETT (Reporter).

WE consider that the most urgent portion of the task intrusted to us is that which concerns the selection and nomenclature of units of force and energy; and under this head we are prepared to offer a definite recommendation.

A more extensive and difficult part of our duty is the selection and nomenclature of electrical and magnetic units. Under this head we are prepared with a definite recommendation as regards selection, but with only an interim recommendation as regards nomenclature.

Special names, if short and suitable, would, in the opinion of a majority of us, be better than the provisional designations "C. G. S. unit of . . . ." Several lists of names have already been suggested; and attentive consideration will be given to any further suggestions which we may receive from persons interested in electrical nomenclature.

The "ohm," as represented by the original standard coil, is approximately 10^9 C. G. S. units of resistance; the "volt" is approximately 10^8 C. G. S. units of electromotive force; and the "farad" is approximately $\frac{1}{10^9}$ of the C. G. S. unit of capacity.

Fig. 8 Examples of naming of electrical units given in 1873 by the British Association for the Advancement of Science, in honour of the respective scientists or engineers (Ohm, Volta, Faraday) [27]

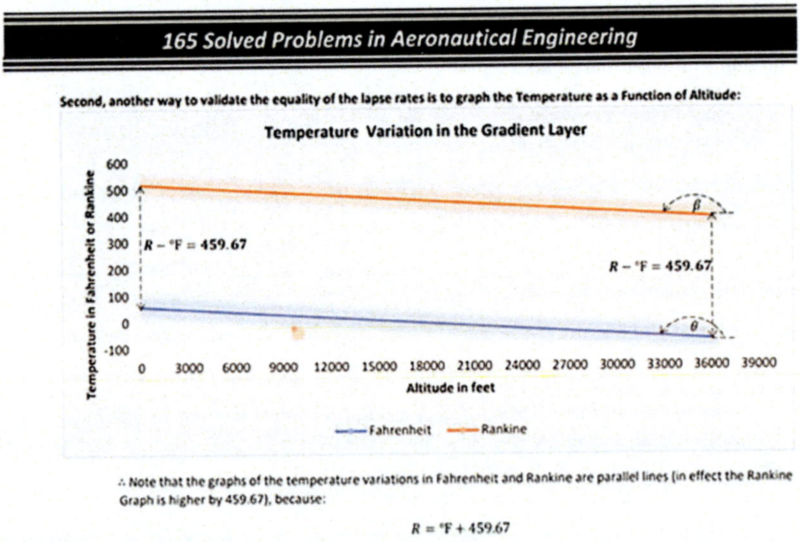

Fig. 9 Aeronautical calculations using the Rankine scale

4 Conclusions

Social perception of engineers and engineering is frequently based on stereotypical conceptions. Literature studies demonstrate, this perception is more critical amongst teachers and students of STEM disciplines. To improve the true integration of STEM education at all educational levels, the development of materials which serve as case studies could be very helpful. In this contribution, the aim is highlighting the E of engineering within the STEM disciplines. Within the frame of the project Engineers and STEM, some videos have been edited promoting the role of engineer's contribution to the development of well-know social, scientific, and technological developments. The videos are designed to be integrated in a full pedagogical approach and will offer a set of relevant characteristics: to show engineer's realizations that wouldn't have been noticed otherwise; to boost intrinsic motivation showing relevance of engineering to the real world; to promote a change of attitude to engineering; and to acquire declarative knowledge about engineering, presenting facts that are a solution reached by a person.

The main structure of the videos, following the above principles, consists of:

(i) Declaration of a well-known fact or product whose name is directly related to an engineer. Presentation of the evidence through an example to draw people's attention to a detail that usually goes unnoticed.
(ii) Asking for the engineer behind this well-known fact or product.
(iii) Presentation of the engineer in short, an explanation of the link between the well-known fact or product and the engineer: how the engineer reached the solution to a practical problem or how his or her contribution was recognized by others so as to name this fact or product with the engineers' name.
(iv) Presentation of some additional information, usually unknown, about the engineer so as to bite the curiosity of the viewer: impact on the society, and on the science and technology system; additional vision of the engineer profile; showing how science, engineering and technology interact.

Links to the URL and screen captures of the videos are given to illustrate the final design. As the project has an international scope, we have developed every video in several languages: English, Spanish, Portuguese, and French. The videos will be allocated in a public website, consisting of a set of exercises, discussion topics, reference documents, and questionnaires on Engineering and Society.

Acknowledgements The authors thanks MERLOT (Multimedia Education Resource for Learning and Online Teaching, www.merlot.org) for the support of the project Engineers and STEM.

References

1. American Society for Engineering Education, https://www.asee.org/. Accessed 20 July 2022
2. Kelley, T.R., Knowles, J.G.: A conceptual framework for integrated STEM education. Int. J. STEM Ed. **3**, 11 (2016)
3. Li, Y., Wang, K., Xiao, Y., Froyd, J. E., Nite, S. B.: Research and trends in STEM education: a systematic analysis of publicly funded projects. Int. J. STEM Ed. **7**, 17 (2020)
4. Leshner, A.I.: Student-centered, modernized graduate STEM education. Science **360**(6392), 969–970 (2018)
5. From the Editor. New STEM and engineering education paradigms. IEEE Control Syst. **38**(5), 8467434 (2018)
6. Sanders, M.: STEM, STEM education, STEMania. Education **68**(4), 20–27 (2009)
7. Simarro, C., Couso, D.: Engineering practices as a framework for STEM education: a proposal based on epistemic nuances. Int J. STEM Ed. **8**, 53 (2021)
8. Cunningham, C.M., Carlsen, W.S.: Precollege engineering education. In: Lederman, N.G., Abell, S.K. (eds.) Handbook of Research on Science Education, pp. 747–758. Routledge (2014)
9. Bucciarelli, L.L.: Engineering Philosophy. Delft University Press (2003)
10. Couso, D., Simarro, C.: STEM education through the epistemological lens: unveiling the challenge of STEM transdisciplinarity. In: Johnson, C.C., Mohr-Schroeder, M.J., Moore, T.J., English, L.D. (eds.) Handbook of Research on STEM Education, pp. 17–28. Taylor and Francis Inc. (2020)
11. Pleasants, J., Olson, J.K.: What is engineering? Elaborating the nature of engineering for K-12 education. Sci. Ed. **103**(1), 145–166 (2019)
12. Kuvac, M., Koc, I.: Enhancing preservice science teachers' perceptions of engineer and engineering through STEM education: a focus on drawings as evidence. Res. Sci. & Tech. Ed. (2022)
13. Kuvac, M., Koc, I.: The effect of problem-based learning on the environmental attitudes of preservice science teachers. Ed. Studies **45**(1), 72–94 (2019)
14. Hammack, R., Utley, J., Ivey, T., High, K.: Elementary teachers' mental images of engineers at work. J. Pre-College Eng. Ed. Res. (J-PEER) **10**(2), 35–46 (2020)
15. Capobianco, B.M., Dieffes-Dux, H.A., Mena, I., Weller, J.: What is an engineer? Implications of elementary school student conceptions for engineering education. J. Eng. Ed. **100**, 304–328 (2011)
16. Fralick, B., Kearn, J., Thompson, S., Lyons, J.: How middle schoolers draw engineers and scientists. J. Sci. Educ. Technol. **18**, 60–73 (2009)
17. Karatas, F.O., Micklos, A., Bodner, G.M.: Sixth-Grade students' views of the nature of engineering and images of engineers. J Sci. Educ. Technol. **20**, 123–135 (2011)
18. Karppinen, P.: Meaningful learning with digital and online videos: theoretical perspectives. AACE J. **13**(3), 233–250 (2005)
19. Schwartz, D. L., Hartman, K.: It's not video anymore: designing digital video for learning and assessment. In: Goldman, R., Pea, R., Barron, B., Derry, S.J. (eds.) Video Research in the Learning Sciences, pp. 335–348. Erlbaum, New York (2007)
20. Project Engineers & STEM. https://www.merlot.org/merlot/viewSite.htm?id=9164223. Accessed 20 July 2022
21. MERLOT (Multimedia Education Resource for Learning and Online Teaching, www.merlot.org). Accessed 20 July 2022
22. Pietri, E.S., Johnson, I.R., Majid, S., Chu, C.: Seeing what's possible: videos are more effective than written portrayals for enhancing the relatability of scientists and promoting black female students' interest in STEM. Sex Roles **84**, 14–33 (2021)
23. Miller, G.: The magical number seven, plus or minus two: some limits on our capacity for processing information. Psychol. Rev. **63**, 81–97 (1956)
24. Moreno, R., Mayer, R.E.: Verbal redundancy in multimedia learning: when reading helps listening. J. Ed. Psychol. **94**(1), 156–163 (2002)

25. Creative Commons (2022). https://creativecommons.org/. Accessed 20 July 2022
26. Universidad de Burgos, Repositorio Institucional RIUBU. https://riubu.ubu.es/handle/10259/2684. Accessed 20 July 2022
27. Biodiversity Heritage Library. https://www.biodiversitylibrary.org. Accessed 20 July 2022
28. Google Books. https://books.google.com. Accessed 20 July 2022

Circulation as a Concept Graphically Represented in a Game-Like Manner to Support the Design Process in Architecture

Evandra Ramos Victorio and Doris Catharine Cornelie Knatz Kowaltowski

Abstract Discussions on topics related to circulation in spaces of the built environment are part of an architectural design process. This exploratory research approaches such topics related to circulation in architecture and how concepts on questions of organization, functionality, hierarchy and orientation can be graphically represented in a visual and standardized manner, in a set of pictographs, to support the design process at the synthesis stage of a project through Design Thinking. A design tool was created, in the form of a card game to support teaching and learning of architecture in design disciplines. In order to assess the efficacy of the tool, classroom activities were created for architecture students. The research presents the transposition of pictographs which include informative narrative and graphic content for the composition of the concept cards and the pathway sheets. The development of our game called *Concept and Idea*, its instruction booklet, as well as the results of the game application are presented here. Our hypothesis is, that the game, due to its informative contents translated in graphic language, may stimulate, still during the design learning process, for a better understanding of the structure of circulation systems in architectural design. As a result, of the applied test, the game activity proved to be favorable as a teaching instrument of those topics related to circulation in architecture in a playful way. The face-to-face teaching activity was considered important, but for future studies one may consider adapting the game to a digital media for application in remote teaching as well.

E. R. Victorio (✉) · D. C. C. K. Kowaltowski
School of Civil Engineering, Architecture and Urban Design, University of Campinas, Campinas, SP, Brazil
e-mail: evandra@robertoleme.com

D. C. C. K. Kowaltowski
e-mail: dkowaltowski@gmail.com

L. F. M. da Silva and A. J. M. Ferreira (eds.), *3rd International Conference on Science and Technology Education 2022*, Proceedings in Engineering Mechanics,
https://doi.org/10.1007/978-3-031-25401-7_4

1 Introduction

This paper is part of a broader research on circulation in architecture and stems from the previous studies by the authors. A discussion on topics of circulation in an architecture project is presented and a supporting teaching tool was developed. The previous study investigated concepts and aspects of circulation of people and goods in architecture that were obtained from theoretical research. Also, on a practical basis the graphical analysis of reference projects, which were represented by free hand-made drawings, were structured in a *Graphic Matrix of Circulation Concepts* (Victório 2019).

In the literature, the topic of circulation is seen as an element to structure and organize conceptually architectural space (Clark and Pause 1996; Unwin 2013). Circulation is as well as a fundamental architectural system that relates movement in space and time (Ching 2008). Circulation is associated with matters of self-guidance and perception of space through the wayfinding[1] concept in architecture (Darken and Peterson 2001; Hunter 2010; Martins and Almeida 2014). The discussion goes further and includes the contemporary concept of flow in architecture and the importance given to human action as well as the notion of flow that includes mobility in the design process (Sola-Morales 2002).

Discussions on the role of circulation in the process of an architecture design project have been present in architecture practices of the Beaux-Arts context, of the School of Fine Arts of Paris, in the 19th Century and remain present in design propositions of the 21st Century. Late in the 19th Century, investigation on movement in architecture was intensified as an essential element for the architectural experience in understanding the composition and form of a building. In a recent study, according to Stickells (2010), movement is related and articulated to programmatic issues that take place in a simultaneous, dynamic way, with continuous variation and interconnected flowing spaces.

The architectural experience is also influenced by changes in modern day society and it reflects transformations of knowledge. New concepts come to light, such as sustainability, which lead towards circular economy; environmental awareness; accessibility; inclusion and a deeper concern with human beings—human centricity—, that are added to digital technologies and the evolution of buildings industry 4.0 and 5.0. Virtual environmental design and communication have an impact in architectural education and activities (Fukuyama 2018; Salama 2008; Serpa and Ferreira 2019).

In architecture, the thinking and decision-making processes are not linear. Also, issues considered are subject to different analysis. There is the understanding that a certain problem has a multi directional characteristic and is submitted to each individual's own understanding, which changes according to time and context, as per cognitive limitations (Lissack 2019). In other words, design is a process of synthesis,

[1] Wayfinding: a cognitive element that permits a person´s ability to perceive, understand and memorize a certain space and/or a way of a mental representation of a given environment, as well as movement as the driving element.

with a normative and creative feature, which demands an organizing origin. Also, a creative design is the result of the process of exploration of an adequate solution for a presented problem. (Cross 2018).

In agreement, Lynch et al. (2019) highlight the need for college students to develop problem solving abilities, creative thinking, communication skills and teamwork, to respond to demands of the present-day economy that is based on fast changes. In order to overcome challenges in the learning processes, students must be active and have a social reflection in cooperation, with authentic tasks towards the development of professional competences. Also, this process includes self and peer assessment. Innovation in teaching and learning of architectural education considered essential can be supported by gaming as a tool that can enhance skills by students (Nicol and Pilling 2005).

In design teaching, thought development is essential for the construction of the learner. Experimentation (modeling, gaming and other similar activities) based on cognitive theories and creativity is essential. Through design learning students can acquire concept knowledge about cognitive processes involved (Oxman and Planning 2004).

Thus, it is important for teaching and for the practice of architecture that issues regarding circulation should be learned, considering that problems in architecture are related to the complexity of flow, mobility, intensification of movement as well as physical and virtual connections with their exchanges. Our research concentrated on the one hand, to understand circulation concepts such as self-guidance, functionality, hierarchy, organization, in design processes and, on the other hand, support the development of new tools for teaching architecture.

The present research aimed the development of a tool modeled as an analogical card game to support the teaching and learning of design and related disciplines in architecture courses. As for specific goals, the research aimed to verify the contribution of graphic representations of circulation elements in architecture through pictographs that were previously developed. These were transferred, as the informative and graphic content for the card game created, to enhance the knowledge and repertoire of students about circulation concepts, besides promoting a reflection about the contribution of such concepts in supporting the design process in the academic and professional realm.

The tool is developed for the synthesis and problem-solving stage of a design process in architecture. It is a board and card game, which includes the graphic and organizational representation of a set of concepts and circulation aspects in the form of a family of pictographs. In order to check the efficacy of the teaching tool, supporting activities were developed for use in the design studio of architecture courses. Our *Concept and Idea* game and the results of experimentations with the game in a teaching environment are presented.

The game allows the association of theoretical concepts of circulation, to a structural axis of a design project, as a game, to assist in the acquisition of knowledge about feasible and innovative solutions to problem-like situations, at the synthesis stage. The game is an active methodology tool that allows simulations of scenarios, innovation and promotes interaction practices. Theoretical topics of organization, self-guidance, hierarchy, functionality and implementation justify the construction of the game. Important design topics were presented to students as a game, so that they would have access to ways of overcoming challenges in an actual design exercise. Design activities require a repertoire and the game aims at contributing to the creative process in a spontaneous and fun way.

2 Games in Education to Support the Design Process

The main features and gains in the use of design games in architecture and in planning lie in the possibility of isolation and simulation of design problems, generation of ideas and building of consensus, as well as performing research and data collection (Pirinen and Tervo 2020). Moreover, participation and interaction among students for the discussion of real-life situations and the resolution of problems are important factors for the learning process.

Games have a pedagogic guideline that is specific and a structure that leads a group process, with a control of variables, that allow the development of thinking skills. Uncertain situations are prevalent in the design process. The stimulate perception and favor separation of specific aspects in a design situation (Brković Dodig and Groat 2019). Criticism towards design games, as they are called, are related to possible biased solutions relating to a specific design, as games have pre-established scopes, parameters and rules (Pirinen and Tervo 2020).

Most design games concentrate on problem solving tasks. As they present goals, requirements, conflicts, rules, gains and losses, and demand an interactive structure, they present the players with challenges. Among the features that determine the player's experience are the requirement of solving a given problem, in a fun way, as well as generating new problems that will promote the continuation of the game (Schell 2008). Those features may be associated in different ways to provide unique experiences in order to motivate and stimulate players (Charsky 2010).

Serious games, or the ones designed for teaching, provide users with teaching and information technology contents, as well as the communication of the game elements, which makes them practical, useful (serious) and playable (Medina et al. 2013). This content aims to integrate fun and learning, so true knowledge can be acquired (Charsky 2010). Academic games, as a methodology in itself, are motivators and can fill in the gap between theory and practice. They also facilitate knowledge in the learning process (Severengiz et al. 2020). A further advantage is that the game dynamics also contribute to reduce passive knowledge and prompts students to make their own decisions when facing complex problems.

Serious games become more efficient as they provide a more flexible structure (Moloney et al. 2017). Due to their features—challenges, rules, narratives—usage of the game favors improvement of cognitive abilities and allows the player to gather theory and practice by using their own previous experiences to make a more personal and attractive learning experience. Games help in the learning process because they make it possible to create experiences that convert into learning and contribute in the improvement of imagination, creativity and language (Ramos et al. 2016).

As pedagogic tools in higher education of architecture, according to the literature, games show efficacy in improving quality and depth of learning and in terms of student's involvement in real life situations through a cooperative and experimental learning process (Brković Dodig and Groat 2019). Serious games are acknowledged as a current tool for teaching because they are capable of intensifying visual and experimental learning as they provide active participation towards construction of ideas and problem solving among students (Álvarez-Rodríguez et al. 2014). Moreover, playing stimulates awareness and improves attention and memory (Batista de Sousa and Miskinis Salgado 2015). Games also create interaction and the development of socio and emotional abilities, such as negotiation, empathy, communication and sportiness.

In games, knowledge passed on to interested parts happens indirectly through direct experience with the game and in open processes that stimulate players to make decisions. A game has the potential to reinforce cooperational exchanges and the capacity to simulate and represent an interaction with the original environment. It can establish alternative goals, imaginable interactions, stimulate learning and questioning (Holland and Roudavski 2016). Games have the capacity to simulate real life circumstances, accelerate the learning process, promote community experiences and have persuasive capacities (Ferrara 2013).

3 *Concept and Idea* Game Description and Justification

For present-day teaching, especially at college level, innovative models of education and active methodologies, including Design Thinking concepts, facilitate dynamic teaching, with more autonomy and common sense, which is necessary for society nowadays. Values such as learning, memory, building on mistakes, efficiency, satisfaction and acceptance are absorbed through gaming (Bittencourt 2017).

Literature shows methods used to develop educational games, also known as serious games, that are adapted for Design Thinking (Bittencourt 2017; Fernandes et al. 2006; Murakami et al. 2014; Sperhacke and Bernardes 2017). Understanding specifications for the development of serious or educational games based on Design Thinking, our development followed steps for the conception of a serious, analogical and purposeful game. These were: objectives and definition on the type of game (defining); creation and development of the game (ideas); development of the playing activities and questionnaires (development); building of game prototypes (prototyping); use of the game, results and necessary adjustment (testing) in order to describe the design process of the *Concept and Idea* game.

We started off with a family of pictographs for circulation concepts in architecture (Victório 2019). First, such concepts were transformed into a scheme, composed by hand made pictures. Then by means of similarity, abstraction and graphic resources, those pictures were converted into pictographs. Later, the pictographs were accompanied by a name and description of the concept—which translate them—to make up the concept and pathway cards that eventually integrate parts of the *Concept and Idea* game.

The game components are: 98 Concept cards; 38 Pathway cards; 08 Opportunity cards; 04 Objective cards and 01 board (as shown at Figs. 1 and 2). Each of the concept cards displays a pictograph accompanied by the name and description of the concept they convey. They are identified by a taxonomy according to a category and structural organization system of each of the six groups of concepts and aspects of circulation (organization, self-guidance, functionality, hierarchy, implementing and compound elements) of a Graphic Matrix, developed previously by Victório (2019).

Pictographs were used to disseminate information, to enable participants to learn circulation concepts and issues in architecture, which are at times difficult to be explained in words. In this sense, we opted for a tool to facilitate the task of players by

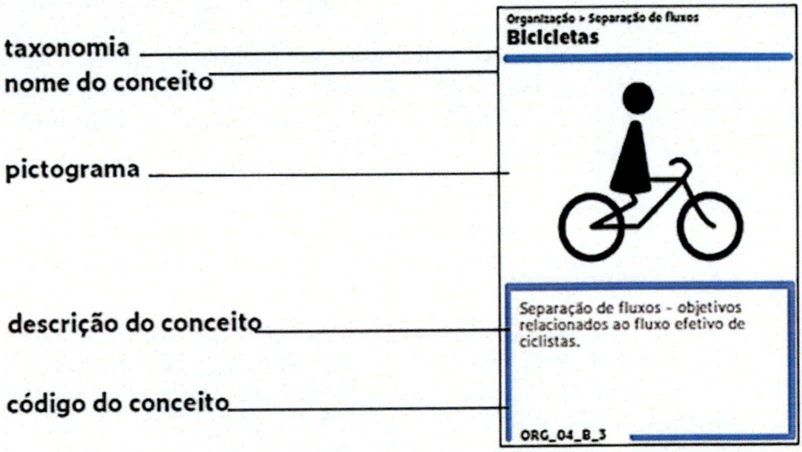

Fig. 1 Example of concept cards

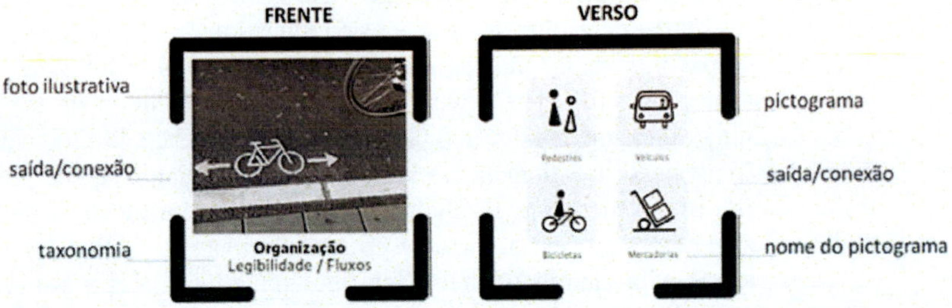

Fig. 2 Example of pathway cards

Fig. 3 Opportunity cards

using graphic images of the circulation elements modeled as pictographs associated with their respective descriptions, aiming at providing structured information.

Pathway cards are made up of the same group of 96 pictographs, in sets of 2 or 4 cards, with information displayed on each side. On the front side, each card shows a picture and either two or four pictographs that resemble the circulation concept shown at the other side of the card. Pathway cards are also classified according to the access setup that is necessary to follow a route within a space, shown on the board, as shown in Fig. 2.

The game also has Opportunity cards that serve as joker cards. They are used to clear the way or to substitute a Pathway card that is already in use in the game, at any point on the board, as per Fig. 3. These cards are presented as a pictograph for said purpose, with name, description, its function and a code. Also, 4 Objective cards were designed to guide the game. Each one of these cards presents a specific objective to be achieved in order to complete the route that links two environments that are represented on the board, on a one-way direction (Fig. 4). For goals specific design objectives were chosen as movement between: museum-garden; lecture room-hall; lab-food court; library-restrooms. A board was created with 26 (blank) slots that allow the development of several routes, as well as 8 environments that are divided into 4 pairs, as shown at Fig. 5. This base was created to facilitate gaming in a rational way.

4 Gaming Description and Activity

In order to validate the tool, the dynamic environment of the game was assembled in an exclusive workshop where the matches were recorded. Each of the five matches was played by 4 students of architecture and urbanism, regardless of the course term

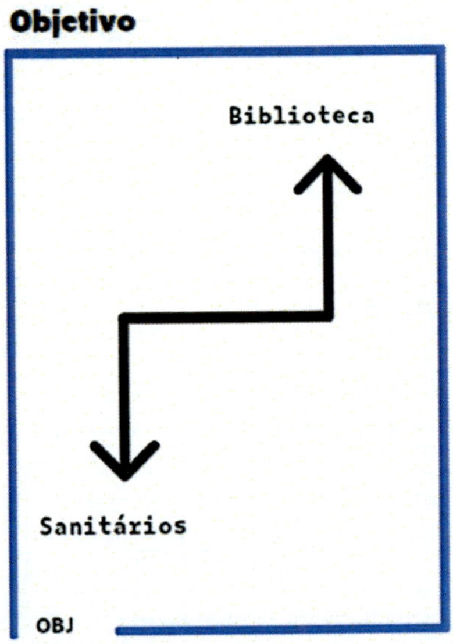

Fig. 4 Example of objective cards

they were enrolled in. There was also a mediator. At first, students were offered a *term of consent and free clarification* (TCLE), game instructions and the digital or printed version of a questionnaire that should be completed at the end of the match. Then, players would gather around the board and received instructions to start the match.

Matches were carried out as follows: each player received 6 Concept cards, 1 Objective card and 2 Opportunity cards, previously described. Upon a dice draw, the first player was appointed and the match would start in clockwise direction. Pathway cards were shuffled and put together with their figures facing up. The first player would collect 3 cards from the top and keep 1 of them according to his/her interpretation and association between the figure and one of the 6 Concept cards. The two remaining cards would be put back at the basis of the card deck. This procedure was carried out by each player, in a sequence. Thus, the route to be completed by each player was determined by the Pathway card chosen at each stage of the match.

The idea behind the *Concept and Idea* game is that players present the Pathway cards to the board in order to complete the route that links the two environments proposed on the Objective card which. This they should do without blocking the way to other players and in such a manner so that the Pathway cards represent a higher quantity of circulation concepts displayed by their respective Concept cards. Thus, the game experience gives the student an association of concepts, elements and aspects of circulation along with their descriptions and representations in a fun and interactive manner.

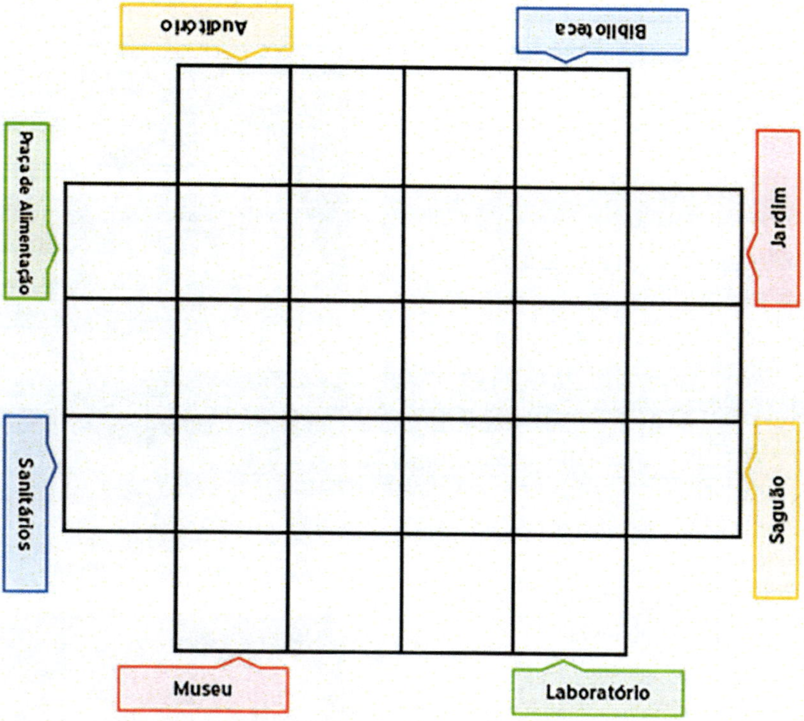

Fig. 5 Board cards

The exits of each chart are connected and they may therefore be rotated. The game continues until one of the players completes his/her route as described by the Objective card. Scores are counted in two moments: when one of the players completes his/her route, that means this player will be the first to create a free route to connect both Objective sites (origin/destination), which grants him/her 2 points. The route should be marked with colorful pieces on the board, as more than one player is allowed to use the same path to reach their destination.

Then, for the second mode of score counting, players turn their Pathway cards with the backsides facing up and with the same position where they were on the board. This will expose the concepts/pictographs of each card, as per Figs. 6, 7 and 8. At this point, each player reveals their Concept cards, which were distributed at the onset of the game and they can compare their cards with the pictographs of the Pathway cards. Players score 1 point for each concept/pictograph on his/her route that is related to one of his/her 6 Concept cards. Finally, the player who scores the most points wins the game. After each match, a discussion and reflection should be carried out concerning the contents presented in the activity and its possible applications in a broader context, as related to an architectural design project.

Fig. 6 Game activity

Fig. 7 Game activity

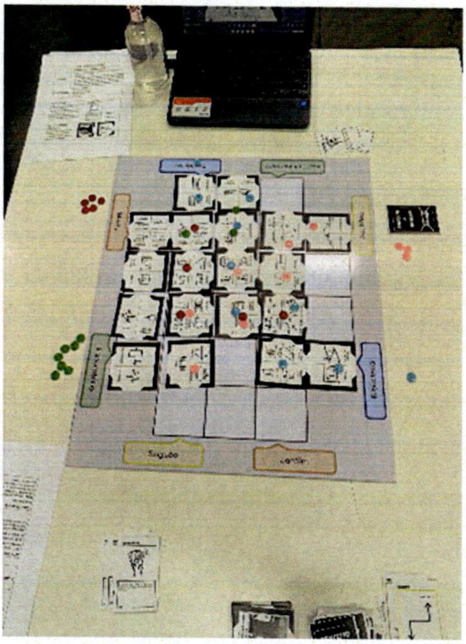

Fig. 8 Game activity

5 Validation of Game Activity

The assessment of gaming results and the game's quality and adequacy as an educational tool (product) was carried out based on a questionnaire answered by 16 students that participated in the matches (20 students participated in the game as volunteers, but 4 of them did not respond the questionnaire). The discussions generated during the game dynamics were analyzed as well.

The questionnaire[2] had 14 questions: 3 close-ended ones within a Likert 5-point scale—(5) Fully agree; (4) Agree; (3) Neutral; (2) Disagree (1) Fully disagree—for quality measurement of the content of the game, its dynamics (playfulness) and Concept cards; 3 assessment multiple choice questions related to the type of discussion that the game prompted. Within the close-ended questions 3 questions with 3 possible answers: yes, maybe, no—were added. The answers to those questions should be followed by justifications. The questionnaire was completed with 5 exploratory open questions related to the game, its dynamics and the learning it provided, which facilitated identification of flaws and chances for improvements to the game.

[2] Available at: <https://forms.gle/MvMje6ToVbQcX45SA>. Accessed on: June 02nd, 2022.

6 Game Results Obtained from the Questionnaire

In the answers assessed according to the Likert scale, 75% of students (12) answered *yes—fully agree* and 25% *agree* (4) to the statement: *The content presented by the Concept and Idea game contributed to your knowledge about circulation concepts in architecture?* Main comments offered about the game were related to learning new concepts, improvement of repertoire and the reasons for its use later; the range of architecture concepts that were presented in such a manner that they can be understood easily and fast; the relationship between images and pictographs with the guidance of routes.

About the game dynamics, 56.3% of students (9) answered *yes, fully agree*; 37.5% (6), *agree*; and 6.2% (1) opted neutral in their answers to the question: *can the game dynamic at Concept and Idea contribute to the structure of a circulation system in your next project?* In their justifications, students considered the game as a method capable of generating discussions about circulation in project and conveying understanding of *multiple choices* of circulation and possibilities of formal design composition and it also helps students in the perception of space notions and provides them with remembered references, terms and circulation concepts and their respective applications, often forgotten in a design process.

The question about the relationship of graphic contents presented by the Concept cards with learning about the topic of circulation—*Did the issues concerning circulation presented graphically (pictograph) in the Concept cards contribute to your understanding and learning about circulation in architecture?* Answers show that 68.8% (11) of students who participated in the game fully agree that there was a contribution to their understanding and learning about the topic, whereas 31.3% (5) simply agree. Handling of the cards as printed material proved to be favorable to understanding a design, as justified by the sentences written by students, obtained from the questionnaire. They also considered it easy to figure out the cards and that they can convey information about circulation concepts *due to the graphic, clear and simple manner in which they are presented*, and also due to the association of terms and pictures and their meanings.

Questions relating to functional aspects of the game:—*What sort of discussion about circulation related issues in architecture does the Concept and Idea game stimulate?*; and *What sort of idea generation and problem solving could be stimulated and/or influenced by the game?* respectively, were answered by 62.5% (10) of students, who stated that the stimulation of ideas concerning issues related to circulation in architecture had something to do with the *practice on how to solve problems in a specific project" and with "functional and technical efficiency*; and 56.3% (9) state that it was related to *comfort issues in architectural design*, as shown at Chart 1 (Fig. 9); whereas 62.5% (10) of answers state that it can prompt the production of ideas about *practices to solve specific design problems*, which could also be a formal questions regarding *solutions of form/volume* and *functional and technical efficiency*; and 56.3% (9) of answers respectively state that the stimulus

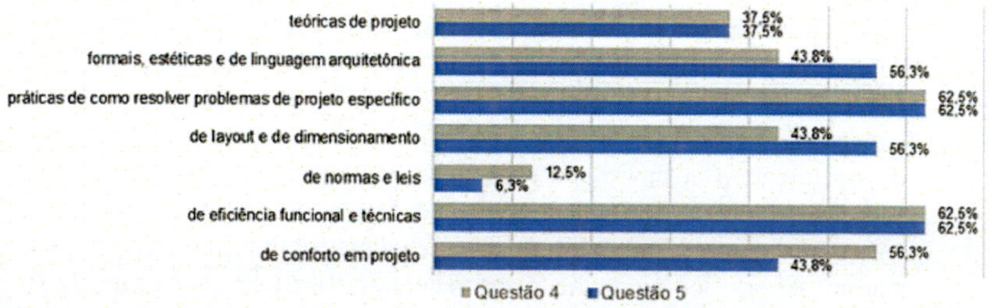

Fig. 9 Chart 1–Answers to questions 4 and 5

might be related to topics of *form, aesthetics and architecture language* and *layout and dimensions.*

The questions that allow three alternative answers (yes, maybe, no) were the following. The answers to: *Does the game have clear and understandable rules?* 87.5% (14) of the students considered it to be true (yes). Only 12.5% (2) answered *maybe.* For questions such as *What did you learn from the game contents?* and *What did you learn from the game dynamics?;* both these questions are open-ended and related to learning acquisition from the game and its dynamics, most students answered that regarding the game content they could learn *modes and types of circulation* and *understanding of techniques and functional efficiency of a design.* They could also experience applicability of circulation concepts for a better development of spaces, access definition and *the flow for a given route.*

Students also noticed that the game dynamics prompted discussion on the interaction among players and their different perspectives regarding circulation. It was clear for them that the practice of identifying and associating circulation topics, that were brought about by the use of Pathway cards, was improved by the game dynamics and that they are related to comfort practices.

For the multiple-choice question: *What did you like the most in the game cards?* which was intended to assess student's preferences, answers such as *rules and playfulness* achieved 62.5% (10 students) of student's preference; and *the graphic part* scored 56.3% (8). *Concept descriptions* were also acknowledged in a 50% (8) ranking among answers. Even though students considered the use of colors a facilitator for perception and classification of concepts, such items were not considered as relevant for the analysis of cards, as per 31.3% (5) of answers, as shown at Chart 2 (Fig. 10).

Fig. 10 Chart 2—Answers to questions 9

Questions related to learning, the game and its dynamics, such as: *With regard to the Concept and Pathway cards, what didn't you like or experienced any difficult with?* And *With regard to the rules of the Concept and Idea game, what didn't you like or experienced any difficult with?* These were open-ended questions intended for identification of flaws, and consequently, to give room for improvement of the game. Most students responded that the game rules were clear and easy to understand. Difficulties reported were related to a better choice of Pathway cards, both in terms of their proper positioning on the board to complete the Objective route, and regarding the association between the Concept card/pictograph and the image on the Pathway card. There were also suggestions that will be taken in consideration for improving the quality of the game, such as reviewing some images for the Pathway cards, the use of the joker card and ranking of players based upon their achieved score. As a result of such findings, adjustment was made in the game to increase the scoring chances from associations.

The question regarding the use of the game in a professional environment—*would you use the game Concept and Idea in a professional environment in group discussions about project?*—81.3% (13) of students responded in an affirmative way, whereas 18.8% (3) answered *maybe*. These affirmative answers given by the participants could be explained by the fact that it is a fun and dynamic activity, which stimulates creativity. They also responded that the activity provides adequate context for discussion and interaction, which is important to understand the ideas generated by the team. They also consider the game an aid to problem-solving in terms of routing due to its logic of organization and layout. For the non-affirmative answers, students considered that the game could be used in the professional context in a shorter version, more focused on the actual design in progress.

Concerning the question about the use of information on the cards in the design process: *would you use the contents of the Concept cards as a basis for the design analysis of your design process?*—68.8% (11) of students answered *yes, because they facilitate understanding types and aspects of circulation that can be created in a design; they enhance repertoire and support problem solving in the design in matters of circulation in a non-obvious way.* 31.3% (5) of students answered *maybe* to this question, but they did not contribute to the last question: *would you have any other comment?* to offer further opinions or suggestions.

7 Conclusion

This article presented a study based on the hypothesis that educational games could be used as supporting tools to a design process in architectural education to support and stimulate practices to enhance discussions about ways to solve design problems and structure circulation systems in architectural design. In order to elaborate on such a hypothesis, research focused on the development of a tool, designed as an analogical game to support teaching and learning of architecture, for design disciplines, specifically related to circulation, and the result of its use in activities carried out in design workshops, especially in the academic context.

The hypothesis could be confirmed by the results the application of the game. It confirmed the initial concept that such experiences are fun, have a graphic content (pictograph), the language of architecture and information concept cards associated with the images. Thus, the experience of the game and its content were beneficial to students. During gaming, interactions between occurred contributing to improve student's knowledge about circulation in architecture. Moreover, answers to the questionnaire also show the feasibility that this tool or game could be used to in general improve the structure of the circulation systems in architecture in future designs. Finally, the game was analogical to permit face to face interactions and increase discussions, however a digital version is possible with on-line applications for distance learning.

References

Álvarez-Rodríguez, F.J., Barajas-Saavedra, A., Muñoz-Arteaga, J.: Serious game design process, study case: sixth grade math. Creat. Educ. **05**(09), 647–656 (2014)

Batista de Sousa, A., Miskinis Salgado, T.D.: Memória, aprendizagem, emoções e inteligência. Revista Liberato, vol. 16, n. 26, pp. 141–152 (2015)

Bittencourt, J.R.: Proposta de metodologia para o ensino e o desenvolvimento de jogos digitais baseada em design thinking a methodology proposal for education and development of games based on design thinking, vol. 20, pp. 1–12 (2017)

Brković Dodig, M., Groat, L.N.: Architecture and urban planning? Game on! In: Brković Dodig, M., Groat, L.N. (eds.) The Routledge Companion to Games in Architecture and Urban Planning Tools for Design, Teaching, and Research, pp. 1–14. Routledge, New York (2019)

Charsky, D.: From edutainment to serious games: a change in the use of game characteristics. Games Cult **5**(2), 177–198 (2010)

Ching, F.D.K.: Arquitetura: forma, espaço e ordem. 2 ed. Bookman, São Paulo (2008)

Clark, R.H., Pause, M.: Precedents in architecture: analytic diagrams, formative ideas, and partis. Wiley, Nova York (1996)

Cross, N.: Developing design as a discipline. J. Eng. Des. **29**(12), 691–708 (2018)

Darken, R.P., Peterson, B.: Spatial orientation, wayfinding, and representation. In: Stanney, K. (ed.) Handbook of Virtual Environment Technology, pp. 353–390 (2001)

Fernandes, K.T., Lucena, M.J.N.R., da Silva Aranha, E.H.: Uma Experiência na Criação de Game Design de Jogos Digitais Educativos a partir do Design Thinking. Novas Tecnologias na Educação **4**(2001), 1–10 (2006)

Ferrara, J.: Games for persuasion: argumentation, procedurality, and the lie of gamification. Games Cult. **8**(4), 289–304 (2013)

Fukuyama, M.: Society 5.0: aiming for a new human-centered society. Japan SPOTLIGHT, n. August, pp. 47–50 (2018)

Holland, A., Roudavski, S.: Design tools and complexity: mobile games and collective imagination. In: Proceedings of the International Conference on Education and Research in Computer Aided Architectural Design in Europe, vol. 2, pp. 555–564 (2016)

Hunter, S.: DR 14—Spatial Orientation, Environmental Perception and Wayfinding. Design Resources, p. 11 (2010)

Lissack, M.: Understanding is a Design Problem: Cognizing from a Designerly Thinking Perspective. Part 2. She Ji, vol. 5, no. 4, pp. 327–342 (2019)
Lynch, M. et al.: Combining technology and entrepreneurial education through design thinking: students' reflections on the learning process. Technol. Forecast. Soc. Change (January 2018), 119–689 (2019)
Martins, L.B., Almeida, M.F.X.M.: O conceito de wayfinding na concepção de projetos arquitetônicos: Interdisciplinaridade a serviço da inclusão. Architecton, vol. 4, n. 6, pp. 1–7 (2014)
Medina, B., Vianna, M., Tanaka, S.: Gamification, Inc: como reinventar empresas a partir de jogos, 1st edn. MJV Press, Rio de Janeiro (2013)
Moloney, J., et al.: Serious games for integral sustainable design: level 1. Proc. Eng. **180**, 1744–1753 (2017)
Murakami, L.C. et al.: Design Thinking como metodologia alternativa para o desenvolvimento de jogos sérios. In: XIX Conferência Internacional sobre Informática na Educação (TISE), vol. 10, pp. 656–661 (2014)
Nicol, D., Pilling, S.: Changing architectural education: towards a new professionalism. [s.l.] Taylor and Francis (2005)
Oxman, R., Planning, T.: Think-maps: teaching design thinking in design education. vol. 25, pp. 63–91 (2004)
Pirinen, A., Tervo, A.: What can we share? A design game for developing the shared spaces in housing. Des. Stud. **69** (2020). https://doi.org/10.1016/j.destud.2020.04.001
Ramos, D.K., Lorenset, C.C., Petri, G.: Jogos Educacionais: Contribuições Da Neurociência À Aprendizagem. Revista X **2**(1), 2016 (2016)
Salama, A.M.: A theory for integrating knowledge in architectural design education. Archnet-IJAR Int. J. Architect. Res. **2**(1), 100–128 (2008)
Schell, J.: The Art of Game Design. [s.l.] Morgan Kaufmann Publishers (2008)
Serpa, S., Ferreira, C.M.: Society 5.0 and sustainability digital innovations: a social process. J. Organ. Cult. **23**(1), 1–14 (2019)
Severengiz, M., Seliger, G., Krüger, J.: Serious game on factory planning for higher education. Proc. Manuf. **43**, 239–246 (2020)
Sola-Morales, I.: Territorios. Barcelona: Editora Gustavo Gili, AS (2002)
Sperhacke, S., Bernardes, M.M.: O processo de ludificação: como transformar métodos de design em jogo de tabuleiro? In: Van Der Linden, J. et al. (Eds.). Design em Pesquisa, vol. 1, pp. 273–300. Marcavisual, Porto Alegre (2017)
Stickells, L.: Conceiving an architecture of movement **14**(1), 41–52 (2010)
Unwin, S.: A Análise da Arquitetura. Routedge, London (2013)
Victório, E.R.: As questões da circulação em arquitetura com base na análise de soluções de projetos contemporâneos. [Universidade Estadual de Campinas]. http://repositorio.unicamp.br/jspui/handle/REPOSIP/333744 (2019)

The Blockchain Technology Applications in Higher Education

Beatriz Vasconcelos, José Luís Reis, Alexandre Sousa, José Paulo Marques dos Santos, and TRUE Project Team

Abstract Blockchain technology is increasingly revolutionizing, not only the way we perform financial transactions, as well as increasingly gaining space in areas that also need transparency and safety. In this paper it will be identified the potential uses of Blockchain in High Education Systems, outlining its advantages and disadvantages, as well as challenges arising from the adoption of this innovative technology. To understand the Blockchain applications in education, an exploratory study was conducted, based on a quantitative methodology, supported by a questionnaire addressed to teachers, workers of academic services, administration staff, students, data protection officers, and IT technicians. This study was based on information associated with the project TRUE—Transparency of Learning Outcomes through Blockchain Technology, the survey respondents were selected from the contacts of the partners of this project. Analyzing the questionnaire results it was verified that the idea and convenience of Blockchain is not known to most respondents, specifically on High Education professionals. Raising awareness to the topic is thus of primary importance, the TRUE Project intends to help that endeavor by giving targeted information about Blockchain in education, and how it can facilitate relations between High Education institutions around Europe and in Erasmus + programs.

Keywords Blockchain technology · High education · TRUE project

B. Vasconcelos · J. L. Reis (✉) · A. Sousa · J. P. M. dos Santos
University of Maia - ISMAI, Avenida Carlos de Oliveira Campos, Maia, Portugal
e-mail: jreis@umaia.pt

B. Vasconcelos
e-mail: A037174@umaia.pt

A. Sousa
e-mail: asousa@umaia.pt

J. P. M. dos Santos
e-mail: jpsantos@umaia.pt

J. L. Reis · A. Sousa · J. P. M. dos Santos · TRUE Project Team
LIACC - University of Porto, FEUP, Rua Dr. Roberto Frias, Porto, Portugal

L. F. M. da Silva and A. J. M. Ferreira (eds.), *3rd International Conference on Science and Technology Education 2022*, Proceedings in Engineering Mechanics,
https://doi.org/10.1007/978-3-031-25401-7_5

1 Introduction

Blockchain is a term that has been gaining traction in recent years. Bitcoin and other cryptocurrency users are probably already familiar with this concept, which refers to a large public record of all transactions made with a given virtual currency.

This paper aims to make known the blockchain technology and its applications in Higher Education, especially, within the context of the work being done in the TRUE project.

The paper begins with an introduction and a brief presentation of blockchain technology, followed by a framework on blockchain in higher education and the advantages of its use in this area of activity. In the following section, the TRUE project is presented. This is then followed by the presentation of the project methodology, with information about the sample and the questionnaire. Next, the analysis of the results and their discussion is presented. Finally, the conclusions of the study are presented.

2 Blockchain Technology

Blockchain has several features. Briefly, it allows information to be stored and distributed reducing the risk of its integrity being compromised. Those who use the blockchain are the same users who keep it up to date, thus creating a "chain" of "block" data that cannot be changed. A "book" and blockchain end up walking side by side, despite being different concepts. What we call "ledger" is where all operations and transactions carried out within the Blockchain are recorded, thus becoming a fundamental piece for attributes such as transparency and security of this technology. This ledger is like a database, in which all Blockchain information is stored. However, this data is distributed on many computers ensuring that this information is difficult and almost impossible to modify or tamper with (Christidis and Devetsikiotis 2016). The structure of the blockchain technology contains blocks, sets of transactions, and pointers to the previous block, as we can see in Fig. 1.

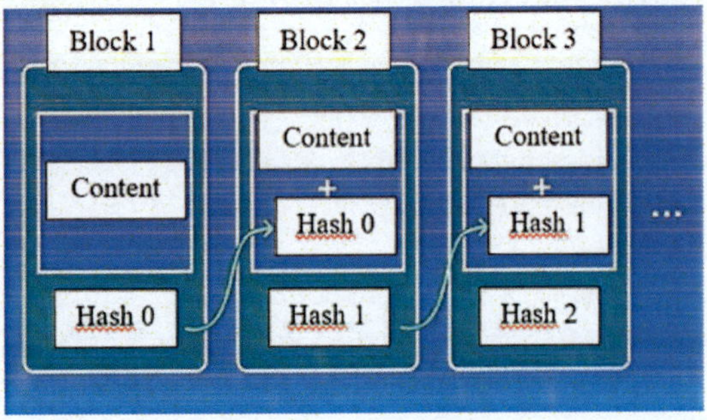

Fig. 1 The blockchain structure—adapted from (Christidis and Devetsikiotis 2016; Palma et al. 2019)

3 Blockchain in Higher Education

Nowadays, higher education has expanded expressively around the world so that, this promising technology has found high-quality applications in this field (Lyceum 2021).

The issuance of transcripts and course completion certificates and the validation of those received from other educational institutions is a frequent need in universities and schools, this represents a significant administrative overhead and there is always the risk of document fraud.

In the educational sector, there are people with different profiles and institutions with peculiar characteristics that enable many use cases for the use of this technology, such as those listed: protection of certificates or diplomas, accreditation of institutions, recognition and transfer of credits, protection of intellectual property, receiving payments from students and financing students through vouchers (Sharma et al. 2020).

Blockchain opens a new approach to Education. Using it, it is possible to reduce administrative costs and most of the work can be automated (Chen et al. 2018). Considering all the benefits of this technology in Higher Education, many universities have started to create their own initiatives and to explore the benefits of applications in the field of education (Grech and Camilleri 2017), as it is the case of the TRUE Project.

4 The TRUE Project

In the context of education in Europe, there is still little standardization of student records. The project TRUE—Transparency of Learning Outcomes through Blockchain Technology, is funded by Erasmus+/KA2 Strategic Partnerships/Higher Education, it aims to develop an integrated training strategy for the assimilation of Blockchain technology in Higher Education Institutions (HEIs) across Europe. To this end, the project produces a basic knowledge pack on the potential of Blockchain technology adapted to Higher Education, including a competency framework for the Blockchain expert (HEI staff member); training content and tools aimed at HEI staff on Blockchain technology, implementation of a MOOC (Massive Open Online Course), as well as a serious game that provides an innovative learning experience on how Blockchain technology works (IPP 2020; ISMAI 2020).

TRUE does not aim to deliver "yet another course" on the basics of Blockchain, rich in high-level theoretical and technical discourse and poor in practical input and scenarios, but to explore how Higher Education Institutions can make it work for them and the resources/opportunities implied using an innovative and comprehensive educational pathway (TRUE Project 2020).

5 Methodology

Based on the objectives of the TRUE project, a quantitative methodology was used, based on a questionnaire, built on a literature review basis. An exploratory study was carried out, in this way, questionnaire respondents, were able to evaluate their understanding of blockchain concepts and their application in higher education.

5.1 *Sample*

The questionnaire was directed to higher education professionals, related to the TRUE project, served to understand how much respondents knew about the blockchain technology, if they were aware of its benefits and the feasibility of its application in higher education, asking specific questions about the blockchain theme. The questionary was intended to be filled by 10 people selected by each partner of the project, in this way were obtained 61 respondents.

5.2 *Questionnaire*

The questionnaire built for this study have 8 sections cover the following: introduction questions, base knowledge, issues, longevity, projects, barriers, Erasmus TRUE Project, comments. Table 1 presents the structure of the questionnaire, and its questions are presented.

6 Result Analysis

The questionnaire was applied to the institutions that participate and are partners of the TRUE project. A total of 61 valid responses were obtained, collected during the months of May and June 2020, and were analyzed. The goal of this activity was to better understand how much and how in-depth the concepts of blockchain and all its related applications have penetrated in the European HEIs.

On the first question, regarding work information, 29% of the respondents are teachers, 23% work in the academic services, other 23% work in administration, 8% are students, only 2% work in the data protection office, 7% are IT technicians, 3% work in the rectory and 5% work in other areas.

To the question 2 "Do you think your organization's information system gives you guarantees of security in all processes in which there is interaction through different communication channels?" the answers are represented in Fig. 2 where it can be observed that for most of the people on this sample, the guarantees of security might

Table 1 Questionnaire structure

Sections	Questions
1. Introduction questions	1. Work information (information about where you work)
	2. Do you think your organization's information system gives you guarantees of security in all processes in which there is interaction through different communication channels?
	3. Do you think your organization's document and workflow management system comply with security rules and that the validation of documents guarantees their reliability?
	4. Country of work
2. Base knowledge	5. How much do you know of: • How a Blockchain works? • How smart contracts work?
	6. Do you know the difference between: • "Public (permissionless) Blockchain network" versus "Private (permissioned) Blockchain network" • "Proof of work" versus "proof of stake"
3. Issues	7. Are you aware of some of the issues and potential problems of Blockchain technology? • Energy consumption (environmental cost) • Eventual need for hard split to fix bugs • Reputational risk (association with cryptocurrency used for crime and tax evasion) • Lack of regulation (risky in an educational setting) • Complexity and potential for becoming slow and cumbersome as the network size increases
	8. Other potential issues: (optional)
4. Longevity	9. Do you think, in the context of higher education, the Blockchain technology is important enough and viable to be here to stay in the long run?
	10. Other comments on long term impact of Blockchain in education: (optional)
5. Projects	11. Do you know any of these sample projects that use Blockchain in an educational or training setting? • BCDiploma (Open Badges compatible academic and professional micro certifications with skills and achievements) • BitDegree (acquire skills required by the job market) • Blockcerts (standard for credentials) • CMF (Common Microcredential Framework) • dAppER (exams distribution and auditability) • EduCents (combat school dropout in Cambodia) • EduCTX (micro credentials) • Kudos (educational effort and reputational reward) • SkillCheck (skill verification for workforce hiring) • Tutellus.io (link education to employment)

(continued)

Table 1 (continued)

Sections	Questions
	12. Additional projects of Blockchain in education or training that you are aware of: (optional)
	13. Do you or did you participate in a Blockchain project? If yes provide some information about it (e.g., a link or a short description of it) (optional)
6. Barriers	14. What do you think are the major barriers or obstacles for using Blockchain and smart contracts in an educational setting? • Awareness • Organizational issues • Governance • Potential security issues due to bugs in the smart contracts' implementation • Deployment • Standardization • Difficulty of choice between competing solutions • Interoperability • Evolution • Environmental impact
	15. Other barriers or obstacles: (optional)
	16. If you were to implement a Blockchain project into a higher education institution, where would you start in terms of priorities (high to low)? • Microcredits • Normal credits • Motivation and reward system • Combating fraud • Auditability
	17. Other priorities: (optional)
7. Erasmus TRUE Project	18. How much importance do you assign to the following 3 intellectual outputs (IO) to be developed within the TRUE Project: • IO1 TRUE Blockchain in Higher Education baseline knowledge • IO2 TRUE MOOC • IO3 TRUE Serious Game
	19. Can you suggest "information units" or subjects that you would like to see covered in the "IO1 TRUE Blockchain in Higher Education baseline knowledge"? (Optional)
8. Comments	20. Global comments: (optional)

be a problem. On the other hand, 18 respondents think that is not a problem at all. Besides these two sides, 4 respondents consider that this issue is a serious problem.

To question 3, "Do you think your organization's document and workflow management system comply with security rules and that the validation of documents guarantees their reliability?". It is conceivable to establish a connection with the answers to

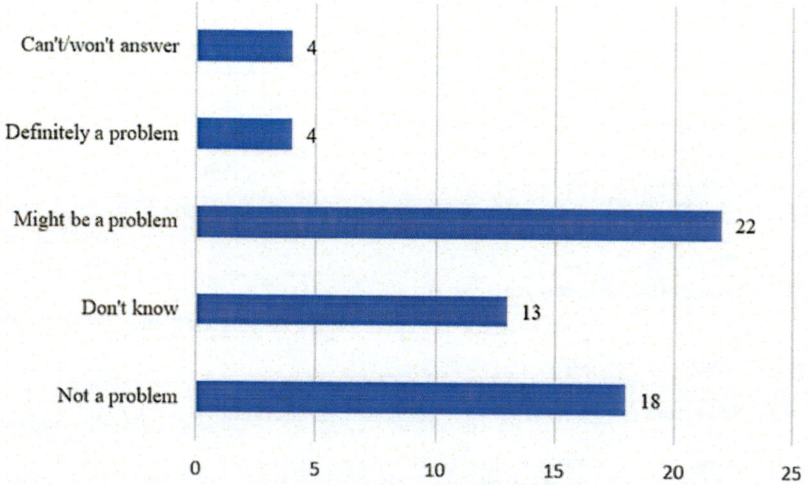

Fig. 2 Information system gives guarantees of security in all processes

question 2, 25 of 61 respondents consider that this is not a problem, while 18 respondents think that might be a problem. Still, there are 3 respondents who consider this question a real and serious problem.

About the country where they work, with 9 options given on question 4 (countries linked to project TRUE, and an option for other countries), 10 respondents answered from Finland, 0 from France, 10 from Germany, 3 from Greece, 3 from Italy, 12 from the Netherlands, 15 from Portugal and 8 respondents from other countries, which are not linked to TRUE Project.

6.1 Descriptive Analysis

After the collection of responses to the questionnaire, it proceeded to the process of the descriptive analysis of the responses obtained.

This questionnaire is divided in 8 groups in total (Base Knowledge, Issues, Longevity, Projects, Barriers, Priorities, Erasmus TRUE Project and Comments).

Regarding the Knowledge Base group and question 5, "How much do you know of how a Blockchain works? How smart contracts work?", we can observe the answers on Table 2, that the greater part of the respondents does not know much about how a Blockchain works, as well as how smart contracts work.

On the question 6, "Do you know the difference between", Table 3 presents the answers about the differences between the two types of blockchains. It is also possible to comprehend that most of the sample do not know the differences between public or private Blockchain networks, nor about "proof of work" and "proof of stake".

Table 2 How much knowledge

	Don't know	Very little	Just the basics	Quite well	Detailed knowledge
How Blockchain works?	10	15	25	7	4
How smart contracts work?	24	3	20	11	3

Table 3 Difference between

	Don't know	Very little	Just the basics	Quite well	Detailed knowledge
"Public (permissionless) blockchain network" versus "Private (permissioned) blockchain network"	23	19	11	4	4
"Proof of work" versus "Proof of stake"	28	17	8	5	3

The **Issues** group of questions aimed to explore the awareness of the target group regarding some of the most common challenges for Blockchain implementation and much wider uptake. Particularly in question 7, "Are you aware of some of the issues and potential problems of blockchain technology?", the answers regarding the awareness of the inquiries about topics that are concerns of blockchain are represented in Table 4.

In question 8, "Other potential issues", an open and optional answer, only seven respondents wrote a commentary. Analyzing them, it is understandable that not many respondents have knowledge to comment the theme:

Table 4 Aware of some of the issues and potential problems of blockchain technology

	Yes	More or less	No
Energy consumption (environmental cost)	20	25	16
Eventual need for hard split to fix bugs	9	24	28
Reputational risk (association with cryptocurrency used for crime and tax evasion)	22	23	16
Lack of regulation (risky in an educational setting)	24	23	14
Complexity and potential for becoming slow and cumbersome as the network size increases	15	22	24

- Knowledge level of staff might be low, and willingness to learn and adapt to new method might be a challenge.
- It's a human behavior to say; I don't know it; I don't trust it. As this blockchain is quite unknown, this could be the case.
- Cost of compliance with already existing systems.
- Governance issues of permissioned blockchains.
- Lack of interoperability.
- Lack of know-how on how it should work when implemented.
- Is based only on assumptions on computational hardness.

In question 9, "Do you think, in the context of Higher Education, the Blockchain technology is important enough and viable to be here to stay in the long run?", Fig. 3 shows that only 10 respondents answered, "definitely yes", which reveals, once again, that the topic is not very known between the academic community, despite all the potential applications and benefits. That is also corroborated by the number of respondents that answer, "don't know enough to make an educated guess". Despite only a small part of the target group know about the blockchain, more than half of the respondents are positive regarding the meaningful impact of this technology in Higher Education.

In question 10, "Other comments on long term impact of blockchain in education", only five respondents gave comments on the impact of blockchain in education, which validate the questions above: many people don't know or understand the impacts of blockchain technology in education.

About the group **Projects**, the results concerning question 11, "Do you know any of these sample projects that use Blockchain in an educational or training setting?", the most popular answer is "no", keeping the premise that the projects and applications of blockchain are not very well known among this sample.

In question 12, "Additional projects of blockchain in education or training that you are aware of", it was received only the following comment/project which is

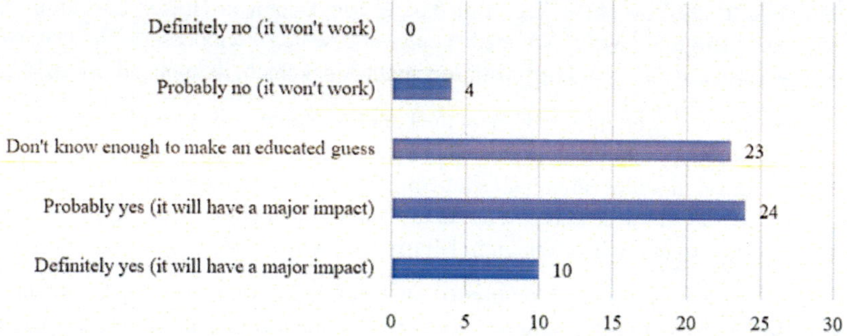

Fig. 3 Context of Higher Education, the blockchain technology is important enough and viable to be here to stay in the long run

"one frictionless experience to exchange trusted career records, using the power of blockchain" (https://www.velocitynetwork.foundation).

Maintaining the tendence, in question 15, "Do you or did you participate in a blockchain project? If yes provide some information about it (e.g., a link or a short description of it)", were received 3 answers, which shows that only 3 respondents, from a universe of 61, participated in a blockchain project.

In question 14, "What do you think are the major barriers or obstacles for using Blockchain and Smart Contracts in an educational setting?" of the group of **Barriers**, whose answers the topics of awareness, organizational issues, governance, potential security issues due to bugs in smart contracts implementation and difficulty of choice between competing solutions, the answers are very similar, and the majority considers that they can be a problem to the use of Blockchain in Education, even though there is a lower percentage that don't know or consider that it is not a problem.

On the group of **Priorities**, question 16, "If you were to implement a blockchain project into a higher education institution, where would you start in terms of priorities (high to low)?", according to the respondents, the greater priority would be combating fraud (high), followed by motivation and reward system (4), auditability and normal credits had the same number (3), normal credits (2) and microcredits (1 low).

In question 17, "Other priorities" were received only 2 answers about some other priorities that should be looked up:

- Getting staff on board.
- Credit transfer.

On group of ERASMUS TRUE Project, in question 18, "How much importance do you assign to the following 3 Intellectual Outputs (IO) to be developed within the TRUE project", the respondents were asked to evaluate the perceived importance of the three IOs. We can conclude that the IO with more importance is IO1—TRUE Blockchain in Higher Education Baseline Knowledge, followed by IO2 TRUE MOOC and finally IO3 TRUE Serious Game.

In question 19, "Can you suggest "information units" or subjects that you would like to see covered in the "IO1 TRUE Blockchain in Higher Education baseline knowledge"?", were obtained 5 answers, which continues to show the lack of confidence and familiarity with and about Blockchain, its applications and baselines:

- Tutorials and information on the benefits and reliability would help.
- Best case applications we can learn from.
- Basic modules on how to start from if we'd want to implement the solution.
- I don't know how any of the above work, thus I gave a "low" for everything.
- How to apply Blockchain in Higher Education.

On the last Group, **Comments**, on the 20nd question, it was asked to the sample for Global comments. Although only 4 respondents left their comments, the common conclusion was that they find this project and Blockchain a very interesting subject and they look forward to knowing more about it. Global comments:

- Very interesting initiative. I would appreciate to learn more on the outcomes. Digitalization during COVID-19 will boost all sorts of digital innovations in the education context.
- Important 'promise' but in 6 years there has been no impact. Standardization is not happening; rules are difficult and many more hindering factors…
- Very much interested in the follow-up.
- There are several projects on blockchain technology, but they are mostly isolated. You should look at the European Blockchain Services Infrastructure and build on that.

7 Result Discussion

The objective of the questionary was to provide a comprehensive overview about the applications of blockchain technology in Higher Education institutions. The TRUE Project partners allowed us to "build" a target-group (Higher Education community) to gain first-hand information about their knowledge on the topic of blockchain.

It is easy to understand that most of the people on the target group is not aware of many of the characteristics of blockchain and the projects that are arising regarding education or other fields.

From the data collected from the questionnaire, we can come to several conclusions: the projects referred in the form are yet very small in scope, so that the general audience is not conscious of them. The fact that there are a lot of applications and only one will not fill all the needs of the institutions may also be a reason for the lack of awareness, as well as the costs, as many institutions do not have the budget to implement such technology.

TRUE could really fill the gaps identified. With a hands-on approach and a comprehensive set of outputs, the project can help raise awareness regarding the best solutions the Blockchain offers for the educational sector. At the same time, it will support the development of important competences that can help HEIs embrace this new technology.

8 Conclusion

With this study, it was possible to obtain a better understanding, related to the technologies used and the development of the TRUE Project, as well to get a sense of the composition and importance of the data kept in an educational environment.

The encryption techniques used in the blockchain ensure security, transparency, longevity, and immutability, making use of a mechanism to eliminate the possibility of data tampering. Given its characteristics, the technology became an object of study for several areas, namely in the educational sector which were addressed in this study, with the aim of reducing human interference in verification and sharing processes of diplomas, as well as creating a scenario, unlike the current one, where in the current methods, there is a need for trust in third parties to guarantee the security of these documents. This work had as general objective to understand a technology with so many benefits and yet so unknown at its core in Higher Education.

The questionnaire goal was to provide a comprehensive view regarding the knowledge and awareness of the blockchain in general and the application of such technology in Higher Education Institutions. The specific objective of the analysis of the questionnaire was to understand the trends, opportunities, and necessities that all this information conceals. Despite the small sample, it was possible to have an insight into the reality: most of the respondents do not know about blockchain technology, the projects carried out on the subject and their application in their area of work, and how they can be used to facilitate it in the long run, despite all the barriers and issues.

The basis of this questionnaire was the TRUE project, which represents a partnership targeting High Education institutions in Europe. The project aims to develop knowledge, give frameworks, training contents and tools to the staff of these institutions to provide an innovative learning experience about how blockchain works. At the same time, it supports the development of competences that can help HEIs embrace this new technology.

To conclude, and as future lines of research, it is possible to understand that despite it not being a subject of common knowledge, it is also probable that it is gaining popularity and conquering many areas, such as education. Blockchain opens a new approach to education, broadening the possibilities of concepts like user data protection, authenticity, security, certificates, and diplomas, as well as others. The decentralization of this technology allows to access valuable information on an independent way and with security. The possibilities of blockchain have not yet reached its full potential, but its core is booming day by day becoming more and more valuable. Because of this, it would be interesting to see, in future works, if more projects like TRUE are implemented and if the educational community becomes more open to this technology. Evaluating the environmental impact of blockchain and comparing with the past, to understand if its benefits overcome the challenges and disadvantages would be of great interest.

References

Chen, G., Xu, B., Lu, M.: Exploring blockchain technology and its potential applications for education. Smart Learn. Environ. **5**, 1 (2018)

Christidis, K., Devetsikiotis, M.: Blockchains and smart contracts for the internet of things. IEEE Access **4**, 2292–2303 (2016)

Grech, A., Camilleri, A.F.: Blockchain in Education. Inamorato dos Santos, A. (ed.) EUR 28778 EN (2017)

IPP.: TRUE PROJECT—Ensino Superior do Futuro https://www.ipp.pt/noticias/true-project (2020). Accessed 20 Sept. 2021

ISMAI.: Transparency of Learning Outcomes through Blockchain Technology (TRUE) https://www.ismai.pt/pt/investigacao/projetos/2020/transparency-of-learning-outcomes-through-blockchain (2020). Accessed 20 Sept. 2021

Lyceum, E.: Blog Lyceum. https://blog.lyceum.com.br/blockchain-na-educacao/#Como_a_tecnologia_pode_ser_aplicada_na_educacao (2021). Accessed 19 Sept. 2021

Palma, L.M., Vigil, M.A., Pereira, F.L., Martina, J.E.: Blockchain and smart contracts for higher education registry in Brazil. Int. J. Netw. Manage **29**(3), e2061 (2019)

Sharma, R.C., Yildirim, H., Kurubacak, G.: Blockchain Technology Applications in Education, IGI Global. Pensilvânia. EUA (2020)

TRUE Project.: Transparency of Learning Outcomes through Blockchain Technology. https://trueproject.eu/ (2020)

Undergraduate Engineering Laboratories: A Study Exploring Laboratory Objectives and Student Experiences at an Irish University

Tom O'Mahony, Martin Hill, and Annie Duffy

Abstract Engineering laboratories are fundamental components in the development of engineering graduates' skills, competences and attributes. However, relatively little existing research has explored the extent to which engineering laboratories meet established laboratory learning outcomes. The contribution of this article is to explore the extent to which laboratory programmes within our Institution are achieving these established Laboratory Learning Objectives (LLO). This was achieved by interviewing 16 engineering instructors across a range of departments. The interviews asked instructors to discuss the aims, objectives, learning outcomes and activities undertaken in a laboratory programme that they were responsible for. The findings suggest relatively weaker evidence in support of LLO 3 (Experiment), 5 (Design), 7 (Creativity) and 9 (Safety). To address these LLO's we argue for the adoption of more open-ended, project/problem/inquiry-based activities in engineering laboratories. The authors are currently collaborating on an Erasmus+ project, Re-OPEN (for more information on the Re-OPEN project, please visit the website https://reopenproject.eu/), focusing on the design of remote laboratories for renewable energies. These findings will be applied to explore ways in which remote laboratories can be designed to be more open-ended and project-based.

Keywords Engineering education · Laboratory practice · Laboratory learning outcomes · Mixed methods

---

Electronic Supplementary Material The online version of this chapter (https://doi.org/10.1007/978-3-031-25401-7_6) contains supplementary material, which is available to authorized users.

T. O'Mahony (✉) · M. Hill · A. Duffy
Department of Electrical & Electronic Engineering, Munster Technological University, Cork, Ireland
e-mail: tom.omahony@mtu.ie

M. Hill
e-mail: martin.hill@mtu.ie

A. Duffy
e-mail: annie.duffy@mtu.ie

L. F. M. da Silva and A. J. M. Ferreira (eds.), *3rd International Conference on Science and Technology Education 2022*, Proceedings in Engineering Mechanics,
https://doi.org/10.1007/978-3-031-25401-7_6

1 Introduction

The role of the engineering laboratory is often implicitly understood to be applying theory to practice or to add some sort of real-world experience to a theoretical programme. However, as Feisel and Rosa observe "since they [laboratory courses] are fundamental to the development of an engineer, learning objectives and their outcomes are critical for evaluating the success and evolution of a laboratory program" (Feisel and Rosa 2005, p. 124). An early attempt to define the role of the engineering laboratory was by Ernst (1983), who "identified three roles or objectives as major ones. First, the student should learn how to be an experimenter. Second, the laboratory can be a place for the student to learn new and developing subject matter. Third, laboratory courses help the student to gain insight and understanding of the real world." This issue was further explored in 2002 by the American Board for Engineering and Technology (ABET)/Sloan Foundation funded colloquium which assembled a group of approximately 50 experienced engineering educators to answer the question "What are the fundamental objectives of engineering instructional laboratories?" The outcome was a set of 13 Laboratory Learning Objectives (LLO) (Feisel and Rosa 2005). Five of these LLO deal with cognition and focus on Instrumentation, Models, Experiments, Data Analysis, and Design. Six were related to broader skills and competences—learn from Failure, Creativity, Safety, Communication, Teamwork, and Ethics in the Laboratory. Finally, two were related to the ability to actually manipulating physical devices, Psychomotor and Sensory Awareness.

Given this set of LLO, the obvious next question is how do we address or achieve these? Felder and Brent (2003) discuss this question, in the context of ABET's criteria for accrediting engineering programmes, and recommended "cooperative learning and problem-based learning, [as] two instructional approaches that have the potential to address all eleven Criterion 3 outcomes effectively". They suggest that perhaps "the most promising approach would be to run fewer but more open-ended experiments. For a given experiment, the students would be given an objective (determine a physical property, establish an empirical correlation, validate or refute a theoretical prediction, ...), provided with enough training to keep them from destroying the equipment or injuring themselves, and turned loose" (Felder and Brent 2003, p. 14). Reports on the status of engineering education contain similar refrains to reform engineering education pedagogy to include more open-ended and project-based learning activities. For example, a Royal Academy of Engineering sponsored report noted that the majority of reform programmes "centre on the implementation of problem-based or project-based learning approach within an authentic, professional engineering context" (Graham 2012). While the 2017 report from the American Society for Engineering Education (ASEE) highlights the need for engineering curricula to include more design-based projects and open-ended problem (ASEE (American Society for Engineering Education) 2017).

However, there is relatively little research exploring the extent to which engineering laboratories are actually addressing or achieving these LLO. Some studies explore this issue in the context of an individual module or innovation. For example,

Oliver and Haim (2009) present an innovate lab-at-home methodology for an introductory digital design module and discuss how the apparatus has the potential to address most, but not all of the LLO. Similarly, but in a mechatronics context, Stark et al. (2013) discuss how their take-home laboratory can address all of the LLO. However, there are fewer studies that explore this issue outside of a single laboratory programme. A notable exception is the study at Virginia Polytechnic Institute and State University in which instructors were interviewed to explore the degree that LLO were achieved (Most and Deisenroth 2003). In all, 19 staff were interviewed from four different departments. The key finding was that "objectives 2 (Models), 3 (Experiment), 5 (Design), 7 (Creativity), and 8 (Psychomotor) received less than convincing ratings, particularly objectives 3 and 5. Judging from informal instructor feedback, objectives 3 and 5 are highly supported as important aspects of the laboratory experience, but time and resources seem to prohibit many of the instructors from addressing these more time-consuming topics" (Most and Deisenroth 2003, p. 9).

The contribution of this article is to then add to this literature and explore the extent to which engineering laboratory programmes within our University are achieving these established Laboratory Learning Outcomes. Given the very limited empirical evidence exploring this topic we believe that this is a significant and important topic that merits further investigation. By exploring this topic, we can help to identify which objectives are regularly met, and those that are less frequently met and consequently make recommendations to enhance the design and implementation of engineering laboratories.

2 Materials and Methods

This study adopted an evaluative, mixed methods approach. Evaluative research is "education research that is conducted to investigate educational programs" (Check and Schutt 2011). An evaluative approach was selected given its close relationship with the focus of this research, which was to evaluate engineering laboratory practice at Munster Technological University relative to an established set of LLO. Evaluative research encompasses a broad school of approaches that can encompass different foci (accreditation, efficiency, effectiveness, outcomes) and schools of thought (experimental, interventionist) (Gray 2004). In this case a goal-based/outcome focused approached was adopted as it best aligned with the particular focus of our research which was on LLO.

Mixed methods combine quantitative methods with qualitative methods and in doing so combine the strengths of both to help overcome the limitations associated with each approach. For example, while surveys allow for the collection and analysis of large quantities of data, the resulting data often lacks insight. In this particular case, the resulting data might tell us the ways in which engineering laboratories were effective but not shed much light on WHY questions—e.g., why were they effective? In contrast, qualitative methods that use interviews typically allow researchers to probe and explore issues in depth. However, they are time consuming to conduct and to analyse, hence it is usually only possible to conduct a small number of individual interviews and consequently it is difficult to generalize from this small number of participants. Adopting a mixed methods approach allows both the what and the why questions to be explored.

In addition to adopting a mixed methods approach, data was collected from two main sources—students and instructors. This enables data triangulation (Fusch et al. 2018). Data triangulation is way of enhancing the trustworthiness of the research findings. For example, if both students and instructors were to identify the same strengths or limitations associated with engineering laboratories then because the data from the two different sources converged independently to reveal the same aspect, we can have greater confidence that this aspect is significant. This makes the finding more trustworthy.

2.1 *Participants*

Our focus was to investigate existing engineering laboratory practices at the module level (rather than programmes or the engineering student body as a whole). Hence our participants were instructors that had responsibility for the design of engineering laboratories and were actively teaching engineering laboratories. In selecting our participants, we tried to be representative and include a range of engineering disciplines, select modules from different stages in the programme (e.g., first year, second year, etc.), strive to achieve a representative gender balance in relation to instructors delivering the modules and, if possible, a variety of modes of delivery (in-person, virtual/remote). 25 faculty were invited to participate and 16 agreed. These participants were asked to identify one module where they were responsible for the laboratory component. Our participants were therefore, these instructors and the students that experienced that selected module. Figures 1 and 2 provide some demographic detail. Figure 1 shows the range of disciplines that were represented. Of the 16 instructor participants, 12 were male and four female. As illustrated in Fig. 2 the majority (75%) of these modules were located either in the first year or fourth year of the engineering programme and the remainder split between second and third year. For the modules that contributed to this study, the average number of practical contact hours per week was 2.9 h.

2.2 *Instruments*

Data was collected from students via an anonymous online questionnaire. It offered the advantage of collecting a larger volume to data and at the same time is relatively fast and convenient for students to engage with. The questionnaire design was largely influenced by the LLO defined by Feisel and Rosa (2005) and the objective was to determine the extent to which STEM students perceived that their laboratory

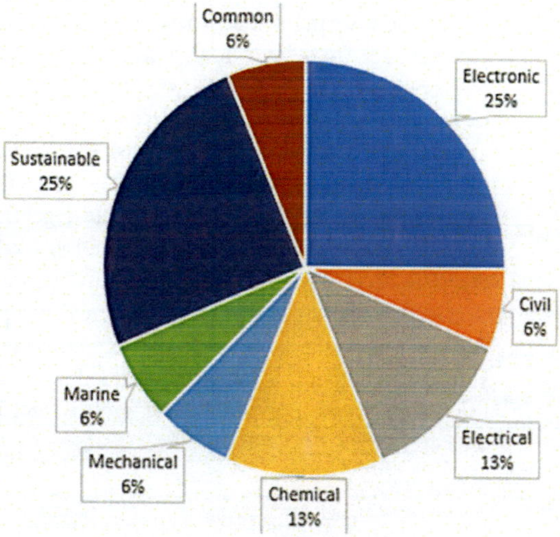

Fig. 1 Disciplines associated with participating faculty members

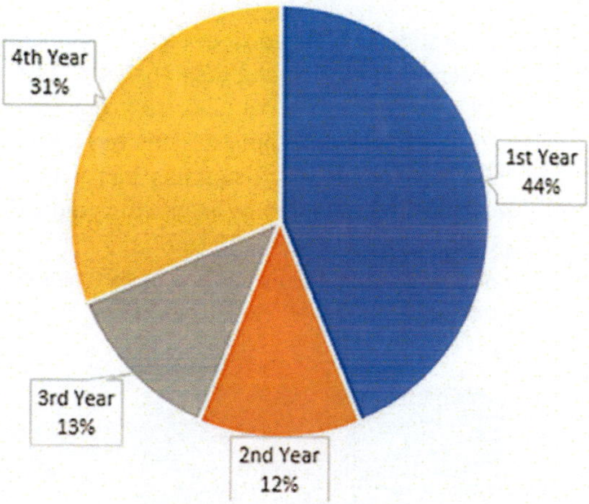

Fig. 2 Participating modules and their corresponding stage in engineering programe

experiences aligned with these established objectives. The rational was that this data would, hopefully, tell us (from the student perspective) which laboratory objectives were being well achieved and perhaps those that were not. To make the questionnaire fast and simple to complete, the majority of the questions were closed-ended Likert-scale response type questions and one open-ended question was included. The one open-ended question asked students to identify the best or most praiseworthy aspect of the laboratory. The questionnaire included a consent form at the beginning and participants explicitly indicated that they consented for their data to be used for this specific research purpose. The questionnaire is available in Appendix 1.

A structured interview protocol was designed to support the gathering of a consistent data set from interviews with individual instructors. Each instructor provided written consent for their data to be used prior to the interview and the interview began by summarizing the aims/objectives of the research and how their data would be used and stored. Participants were then asked to confirm that they understood the nature of the research and provided consent. The interview proceeded by inviting participants to describe the context of the module e.g., the programme that it was associated with, the number of students that take the module, the time spent in the laboratory etc. The second phase of the interview then explored the design and implementation of the laboratory from a pedagogical perspective focusing on the overall aim of the laboratory, the laboratory learning outcomes, the type of activity undertaken by students, whether it is more open-ended or closed-ended, whether students complete the laboratory as individuals or in groups, etc. The average duration of the interview was 40 min.

2.3 *Data Analysis*

The oral data collected from these participants was transcribed using an automated transcription service and then analysed. The overall analytical framework is illustrated in Fig. 3. The 13 laboratory objectives defined by Feisel and Rosa (2005) were selected as an analysis framework as they represent a set of established STEM laboratory objectives. Using this framework, it is possible to analyse the pedagogical approach adopted by instructors and experienced by students as summarized by Fig. 3. Likewise, the framework may support the identification of specific or noteworthy case-studies of good practice that may inform the design of engineering laboratories.

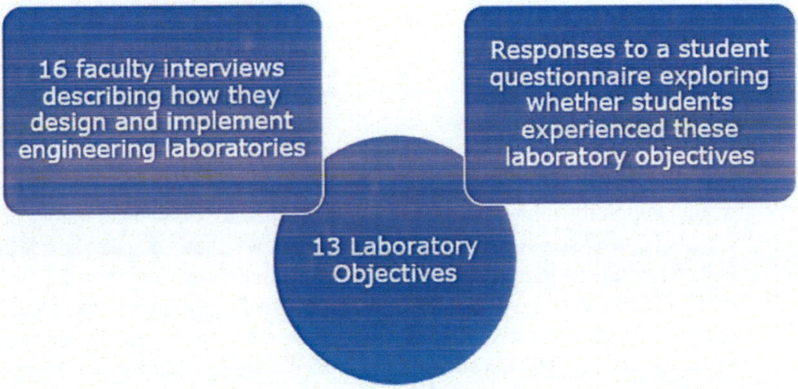

Fig. 3 Overall analytical framework

3 Findings

Table 1 summarizes the overall findings. This table combines the findings from the instructor interviews with responses to the student questionnaire to evidence the achievement of each of the laboratory learning outcomes. Here we have classified "evidence the achievement" rather broadly into one of three categories. A rating of 3 "✓✓✓" indicates the presence of good evidence to support this learning outcome, 2 some evidence and 1 "✓" weak or little evidence. The following two subsections discuss this classification in more detail. It should be noted that the response rate to the student questionnaire was low, comprising 32 responses. Hence this data should be interpreted with some caution.

3.1 *Findings from Instructor Interviews*

Some key findings from the instructor interviews are presented in Figs. 4a–d and 5. Participants were asked to describe the learning outcomes associated with their laboratory programme and these were subsequently classified as being technically focused or relating to broader skills and competences. Many instructors, for example, talked about how the primary objective or outcome was to apply theory to practice or that the laboratory objective was to help students understand the theory. Some instructors said that key learning objectives were related to generic skills such as teamwork, analysis or independent learning. Examples such as these were classified as broad learning outcomes (LOs). Figure 4a reveals that only 25% of participants identified these broader learning outcomes as being key objectives associated with their modules while 50% of participants were designing and implementing modules

Table 1 Triangulating data from university instructors, students and employers with respect to the 13 Laboratory Learning Objectives where "✓✓✓" indicates that there is good evidence to support this learning objective while the presence of a "✓" indicates little evidence to support the learning objective

LLO	Instructor data	Student data
No 1 Instrumentation	✓✓✓	✓✓✓
No 2 Models	✓✓✓	✓✓
No 3 Devise experiment	✓	✓✓
No 4 Data analysis	✓✓	✓✓✓
No 5 Design	✓✓	✓
No 6 Learn from failure	✓✓✓	✓✓
No 7 Creativity	✓	✓
No 8 Psychomotor	✓✓	✓✓✓
No 9 Safety	✓✓	✓
No 10 Communication	✓✓✓	✓✓
No 11 Teamwork	✓✓	✓✓
No 12 Ethics in the lab	✓	✓✓✓

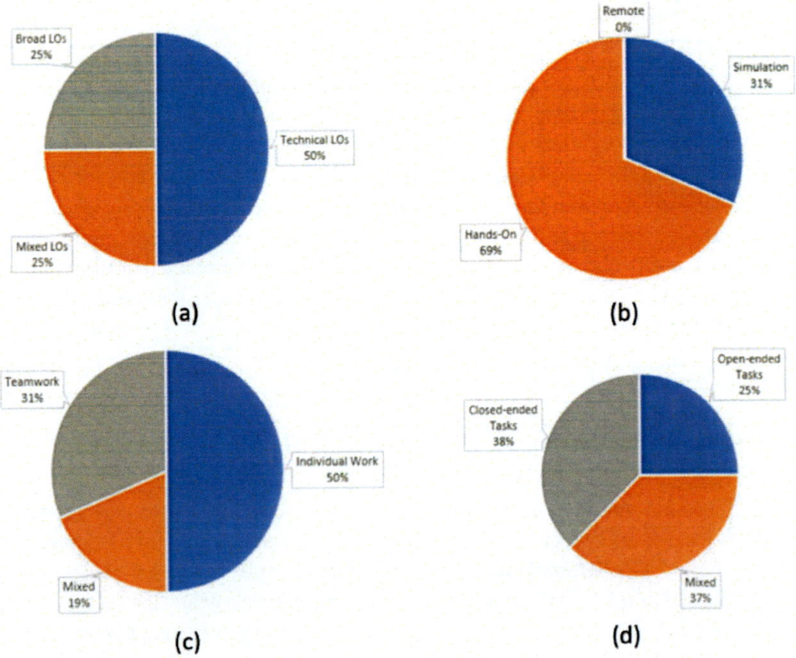

Fig. 4 Summary of key findings from instructor interviews, **a** categorization of laboratory learning outcomes as either focused on technical learning outcomes or broader skills and competences; **b** laboratory mode of access; **c** how students completed the laboratory experiments; **d** categorization of laboratory experiments as either closed-ended problems/tasks or more open-ended problems/projects

where the LO's were predominantly focused on technical achievements. Linked to this, Fig. 4c and d reveal that 50% of participants design modules where the experimental work is completed by individuals and that 25% of instructors design modules where the experimental work is more open-ended (more than one solution is possible) in nature. Linking back to Table 1, these findings suggest weaker evidence to support LLO 3, 5 and 7. LLO 5 and 7 which relate to Creativity and Design are likely to require more open-ended tasks and the infrequency of those types of tasks would suggest that, overall, these LLO are less likely to be met. Given that 50% of the participants designed laboratories where the learning outcomes were technical in nature and explored how theory applied to practice there is good evidence to support LLO 1 and 2. The dominance of hands-on or in-person laboratories, Fig. 4b, generates good evidence for LLO 1: Instrumentation and 8 Psychomotor.

As part of the interview process, participants were asked to name the generic skills (as many as they liked) that were emphasized in the laboratory work that they designed. As is evident from Fig. 5, the most frequently identified skills are problem-solving, technical communication, teamwork, interpersonal communication, health

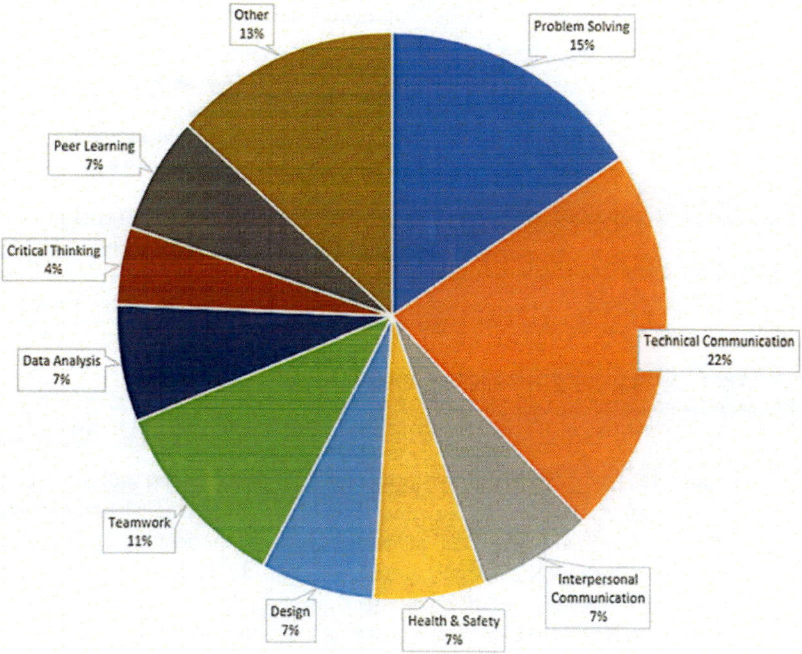

Fig. 5 Key or generic skills most frequently identified by individual instructors

and safety, design and data analysis. Hence, Fig. 5 provides good evidence that LLO 6 Learning from Failure (via problem solving) and Objective 10 Communication are being regularly addressed. While Fig. 4 would suggest that there is relatively little evidence to support LLO 5: Design, as the percentage of modules that include open-ended tasks is low (25%), the fact that design was recognized as one of the top seven key skills generates some evidence to support the achievement of this LLO. Similarly, while only 31% of the modules designed include a formal practical teamwork element, these participants identified teamwork as one of the top seven key skills. Consequently, we attributed a rating of 2 to LLO 5 and 11. Because Health and Safety and Data Analysis were also identified, there was some evidence to support the corresponding LLO and these received a rating of 2. LLO 12: Ethics in the Laboratory did not feature significantly in the interview data.

3.2 Findings from Student Questionnaire

While all students that experienced the modules designed by the participating instructors were invited to participate, only 32 did so. The closed-ended or Likert-scale questions related to LLO were analysed by computing the mean and that data is

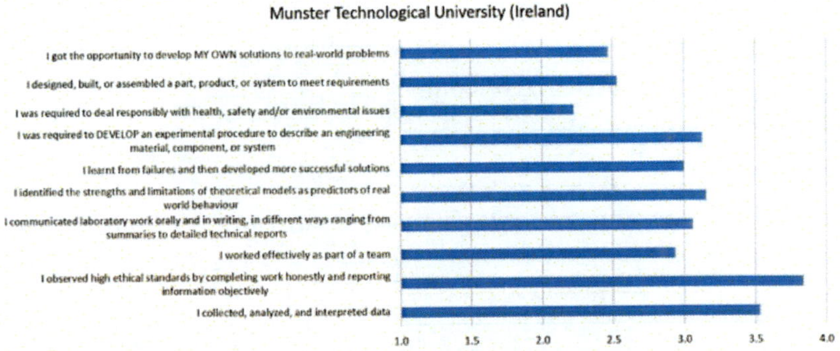

Fig. 6 Average student rating of how frequently they engaged with each of the LLO where 1 = never, 2 = occasionally, 3 = sometimes, 4 = frequently

presented in Fig. 6. This data reveals that these students report regularly experiencing LLO 4 (Data Analysis) and 12 (Ethics in the Lab) during their laboratory practices. From Fig. 6 it is also clear that students reported experiencing LLO 7 (Creativity i.e., developing OWN solutions), 5 (Design) and 9 (Health and Safety) much less frequently. The analysis of the open-ended question, summarized in Table 2, revealed two themes relevant to Laboratory Learning Outcomes, namely the hands-on experience and opportunities to develop specific skills. These themes provide good support for LLO 8 (Psychomotor).

4 Discussion, Implications and Conclusion

Table 1 indicates that there was reasonably good agreement between the two sources of data that we gathered which helps to support the trustworthiness of the research findings. The notable exception to this statement is the evidence gathered in support for LLO 12 (Ethics in the Laboratory). The fact that this question received the highest rating by students in the questionnaire data would suggest that this learning objective is emphasized, at least in the assessment component when reporting and writing up results. However, there may also be a positive bias here and that students "felt" that they needed to respond positively to this question as it is perhaps the only question that directly impacts on their assessment and grades. Interviews or focus groups with students would be useful to explore this further to understand why this question received the most positive rating. In order to minimize introducing bias into the data, we did not explicitly ask instructors about "Ethics in the Laboratory". In so far as it was possible we tried to avoid mentioning any of the LLO. The exception was Objective 11: Teamwork where instructors were asked if the laboratory practice they designed were undertaken by students individually or as part of a team. So, it may be the case that instructors are concerned with ethics in the laboratory, but that it didn't

Table 2 Praiseworthy aspects of laboratory experience as identified by students through the open-ended question

Category	Frequency	Examples
Learning support	13	"The feedback from the lecturer when I had a problem was possibly the best experience of the module" "Being able to watch the lab tutorial multiple times" "Canvas recordings of labs and a beginning excel file to work through the labs at my own pace" "…I learned from my peers as much as they learned from me"
Hands on/Practical experience	7	"Using the equipment and resources to provide a solution to a real-world problem" "The practical hands on experience with the equipment was very helpful to my understanding" "I got hands on experience with software, which is more effective than watching others do it" "The practical lab experiments"
Specific practical skills	6	"Interpreting data and how to conduct statistical analysis were the most beneficial parts" "The way the labs were delivered allowed steady development of information and engineering testing methods" "Working on our projects really tested our Excel skills" "The understanding of how applications can be programmed to do different things"

occur to instructors during the interviews. Or perhaps because it is often linked with assessment, instructors might have felt that it was not that relevant. Never-the-less, the fact that ethical concerns were largely missing from the interview data would suggest that it is less of a concern or a lower priority than other LLO.

Existing laboratory practices at MTU focus on data analysis, using models, interacting with instruments, learning from failure, developing psychomotor skills, communication skills and ethical responsibilities—especially around gathering and reporting data. Existing laboratory practices tend to be designed to encourage students to work as individuals solving predominantly closed-ended problems. Hence, and as documented in Table 1, there was generally a reasonable to good quality of evidence to support the claim that the laboratories designed by these participants addressed seven of the LLO. The ones most clearly addressed are Laboratory Learning Objective 1 (Instrumentation), 2 (Models), 4 (Data Analysis), 6 (Learn from Failure), 8 (Psychomotor), 10 (Communication) and 11 (Teamwork). The evidence in support of LLO 3 (Experiment, especially the element devise an experimental approach), 5 (Design, especially the aspect related to design rather than build or assemble), 7 (Creativity), 9 (Safety) and, possibly, 11 (Ethics in the Laboratory) is weaker. Consequently, we would suggest that more focus needs to be placed on achieving these four or five (if Ethics is included) LLO. The study conducted by Most and Deisenroth

(2003) identified weaker evidence in support for LLO 2, 3, 5, 7 and 8. The identification of a common set of LLO, namely 3, 5, 7 would indicate that this might be problematic for many engineering programmes and a focus of international attention. The outliers might reflect the missions or distinctiveness of different universities and something to be addressed more locally.

It is perhaps not surprising that there is less evidence to support LLO 3, 5 and 7 as, arguably, these are more challenging to achieve. The outline description associated with Laboratory Learning Objective 3 includes statements like being able to "Devise an experimental approach, specify appropriate equipment and procedures" (Feisel and Rosa 2005). Laboratory Learning Objective 5 states that students should be able to "Design, build, or assemble a part, product, or system, including using specific methodologies, equipment, or materials; meeting client requirements; developing system specifications from requirements…." (Feisel and Rosa 2005). Laboratory Learning Objective 7 relates to student's ability to "Demonstrate appropriate levels of independent thought, creativity, and capability in real-world problem solving". As evidenced by Fig. 4, the majority of these participants are designing closed-ended laboratories with an exclusive focus on technical learning outcomes which tends to work against Objectives 3, 5 and 7. These practices should generally be collaborative in nature. Health, safety and environmental issues also need to feature more prominently in more laboratory practices. Given the broad nature of LLO 3, 5, and 7, it is likely that the laboratory experience will need to be re-designed so that it is more open-ended, enquiry orientated or organized around problems or projects and the focus of laboratory learning outcomes shift from just demonstrating theory to have a broader focus on key skills such as design and independent learning. This recommendation would be well supported by the existing literature (ASEE 2017; Felder and Brent 2003; Graham 2012). Considering that, as graduates, many will find themselves working in teams on complex, open-ended problems, more laboratory practices need to pivot and require students to design experiments and require students to design systems that address more open-ended problems. A possible implication is that Universities may need to develop suitable resources and support mechanisms to equip engineering instructors with the competence and skill set to design, implement and support more open-ended, collaborative, enquiry, problem and project-based laboratory experiences.

To this end the Dept. of Electrical and Electronic Eng. At MTU is collaborating on an EU project funded through the Erasmus+ scheme to develop Remote laboratories for Practical Experiments on renewable energies at EU universities (RE-OPEN). The other project partners are Universita Degli Studi Guglielmo Marconi (Italy), Fachhochschule Technikum Wien (Austria), Universidad De Huelva (Spain), Norges Teknisk-Naturvitenskapelige Universitet (Norway) and VJ Technology (Italy). The RE-OPEN[1] project aims to design, test and validate an innovative standard educational model and ICT solution for remote labs—with a focus on renewable energies. This pedagogical model and ICT solution can then be reused at other institutions

[1] For more details on the Re-OPEN project, please visit the website https://reopenproject.eu/.

to reduce the challenge associated with developing remote laboratories. A particular focus will be to examine how remote laboratories can be more open-ended and become more collaborative and problem based. Hence the Re-OPEN project may enable the creation of resources that will support the professional development of engineering instructors to design, implement and support laboratory experiences that specifically address LLO 3, 5 and 7.

References

ASEE (American Society for Engineering Education): Transforming Undergraduate Education in Engineering Phase II: Insights from Tomorrow's Engineers (2017)

Check, J., Schutt, R.K.: Research Methods in Education. Sage Publications Ltd (2011)

Ernst, E.W.: A new role for the undergraduate engineering laboratory. IEEE Trans. Educ. **26**, 49–51 (1983). https://doi.org/10.1109/TE.1983.4321598

Feisel, L.D., Rosa, A.J.: The role of the laboratory in Undergraduate Engineering Education. J. Eng. Educ. **94**, 121–130 (2005). https://doi.org/10.1002/j.2168-9830.2005.tb00833.x

Felder, R.M., Brent, R.: Designing and teaching courses to satisfy the ABET engineering criteria. J. Eng. Educ. **92**, 7–25 (2003). https://doi.org/10.1002/j.2168-9830.2003.tb00734.x

Fusch, P., Fusch, G.E., Ness, L.R.: Denzin's paradigm shift: revisiting triangulation. J. Soc. Chang. **10**, 19–32 (2018). https://doi.org/10.5590/JOSC.2018.10.1.02

Graham, R.: Achieving excellence in engineering education: the ingredients of successful change (2012)

Gray, D.E.: Doing Research in the Real World. Sage, London, Thousands Oaks, New Delhi, pp. 1–570 (2004). https://doi.org/10.1007/s13398-014-0173-7.2

Most, K.R., Deisenroth, M.P.: ABET and engineering laboratory learning objectives: A study at Virginia tech. ASEE Annu. Conf. Proc. 1227–1246 (2003)

Oliver, J.P., Haim, F.: Lab at home: hardware kits for a digital design lab. IEEE Trans. Educ. **52**, 46–51 (2009). https://doi.org/10.1109/TE.2008.917191

Stark, B., Li, Z., Smith, B., Chen, Y.: Take-Home mechatronics control labs: a low-cost personal solution and educational assessment. In: ASME 2013 International Design Technical Conferenes, pp. 1–9 (2013)

Teaching Adhesive Bonding in Mechanical Engineering Courses

A. Q. Barbosa, E. A. S. Marques, R. J. C. Carbas, and L. F. M. da Silva

Abstract Adhesive bonding is widely implemented in many industries, such as the aerospace, automotive, shipping and railway sectors. The increasing popularity of this technology is linked to the noteworthy benefits related to its application, compared to other traditional joining process, such as welding or mechanical fastening (Kurfess, Producing the modern engineer 19(1):118–123, 2003 [1]). An important factor driving this change is the current European Union climate and energy policy, which has established a target of improving energy efficiency in the European Union by 20% by 2020. To meet these targets, multiple industrial sectors are seeking lighter, stronger, more durable and environmentally friendly multi-material structures, which in practice can only be achieved with adhesive bonding. As the usage of this technique has increased, so has increased the necessity to train qualified professionals, including a larger number of newly qualified engineers with skills in this joining technology (Borges et al., Women in mechanical engineering: a case study of the faculty of engineering of the University of Porto in the last 20 years, 2(1):48–67, 2022 [2]). Adhesive bonding is a truly multidisciplinary field, requiring mastery of fundamental concepts from various scientific disciplines (physics, chemistry and mechanics). This study is focused on understanding the way this technology is approached and taught in Mechanical Engineering courses, comparing different methodologies, pedagogical strategies and syllabus contents. Countries which are members of EURADH were considered in this study. EURADH is a European association which aims to join several national associations dedicated to adhesion technologies, and which aims

A. Q. Barbosa (✉) · E. A. S. Marques · R. J. C. Carbas
INEGI, Rua Dr. Roberto Frias, 400, 4200-465 Porto, Portugal
e-mail: aqueiros@inegi.up.pt

E. A. S. Marques
e-mail: emarques@fe.up.pt

R. J. C. Carbas
e-mail: carbas@fe.up.pt

L. F. M. da Silva
Department of Mechanical Engineering, Faculty of Engineering, University of Porto, Rua Dr. Roberto Frias, 4200-465 Porto, Portugal
e-mail: lucas@fe.up.pt

L. F. M. da Silva and A. J. M. Ferreira (eds.), *3rd International Conference on Science and Technology Education 2022*, Proceedings in Engineering Mechanics,
https://doi.org/10.1007/978-3-031-25401-7_7

to disseminate knowledge not only by the scientific community but also to ensure the transfer of knowledge with the industrial community. The top five Mechanical Engineering courses (according to QS World University Rankings, for 2020) in Germany, Italy, the Netherlands, Portugal and the United Kingdom were studied.

Keywords Adhesive bonding · Mechanical engineering · University rankings · Knowledge transfer

1 Introduction

With the evolving occupational training paradigm, especially at the beginning of 21th century, it has been observed that the current requirements of a mechanical engineer are vastly different from those in place just 40 years ago. Kurfess [1] introduced the concept of a "Modern Engineer", which should perform a variety of tasks in its professional role. Consequently, the curricula of mechanical engineering courses should follow this trend and engineering programmes now endeavour to provide students with the skills required to launch successful technical careers. Nowadays, careers in STEM (Science, Technology, Engineering e Mathematics) fields, such mechanical engineering, rank among the fastest-growing [2, 3]. Transversely to all engineering branches, degree programmes should first and foremost provide students with a general education that assists them with the analytical and critical thinking tools [3]. In addition to teaching fundamental engineering concepts, the programmes should also include the most commonly used technologies and those known to be emerging in the industrial sector.

In the last two decades, manufacturing technologies have advanced at a swift rate. Materials are increasingly lighter and stronger and, concurrently, the techniques suitable for joining these materials have also undergone several changes [4–6]. The introduction of non-metallic materials, such as plastics or composites coupled with the necessity of joining dissimilar materials [7], have pushed the industry to apply alternative joining technologies. Modern adhesive and sealant bonding processes are being widely implemented in many technological industrial sectors, which include the aerospace [8], automotive [9], marine [10], railway, footwear [11], electronic components and sports equipment sectors [7, 12]. Ultimately, adhesive bonding is seen as a key asset in the technological toolbox that drives manufacturing to modern, twenty-first century practices [13]. This technique is growing rapidly both in Europe and worldwide, as it constitutes a powerful competitive advantage for companies and industries that can also bring together the knowledge, skills and techniques required to successfully use it. However, this is especially true as there is a growing demand for EU companies to become more competitive through upskilling and modernisation processes [13, 14].

In light of this scenario, the key question posed in this research work is how universities, and specifically the Mechanical Engineering courses, can keep pace with this technological evolution and avoid stagnation. Are the new engineers who are being graduated equipped with the skills and competences demanded by the industry?

In this context, it is fundamental to understand if the curricula of mechanical engineering courses in Europe are following these technological advances regarding adhesive bonding. To ensure the high quality of engineering degree programmes and to keep up with industrial sector training needs, periodic reviews and enhancement of the curricula should be carried out. This task is far from being simple, being challenging on a multitude of levels and those involved in reviewing and improving the curriculum must have a holistic view of current and anticipated needs in the years following graduation. According to Carew and Copper [15] this type of review and update is supported by three drivers:

- keeping pace with the fast technology evolution;
- shifting social expectations and aligned shift with regulation and legislation;
- changing expectations of higher education regulators and stakeholders (e.g., students, academics, government and accrediting bodies).

When revising a curriculum to introduce a new technology, an additional question that arises is: Does my institution have specialised skilled human resources in this subject area?

Typically, the existence of highly qualified personnel in a given scientific and technological area derives, in a large part, from research groups. These research groups, mostly integrated in universities, will support the specialized personnel requirements in a certain thematic subject area [16]. Technology transfer from research groups to graduates and then to industry is thus fundamental for the evolution of science and technology based fields of knowledge [13, 17]. According to Dubickis [18], innovation is also the essential driver for economic growth, generating innovative knowledge at the service of the tangible needs of the industrial evolution. It is therefore of utmost importance to ensure that research groups associated with universities can transfer knowledge and assist universities by providing qualified personnel to lecture on emergent technology.

This study aims to understand how adhesive bonding is approached and taught in Mechanical Engineering courses, doing so by comparing different methodologies, pedagogical strategies and syllabus contents. Countries which are members of EURADH were considered in this study and the top five Mechanical Engineering courses (according to QS World University Rankings, for 2020) in Germany, Italy, the Netherlands, Spain, Portugal and the UK were the subject of study.

2 Methodology

2.1 Selection of the Countries

For this study, countries with representation on EURADH were selected. EURADH holds an international conference every two years in Europe which aims at gathering researchers from University, Industry and Technical Centre's to present and discuss the current state of progress in the field of adhesion. The scientific committee of this conference in 2020 consisted of representatives from seven European countries:

- Portugal was represented by Associação Portuguesa de Adesão e Adesivos (APAA);
- Spain was represented by Grupo Español de Adhesion y Adhesivos;
- France was represented by French adhesion division (SFA) of the French Vacuum Society (SFV);
- United Kingdom was represented by Society of Adhesion and Adhesives;
- The Nederland's was represented by De Bond voor Materialenkennis:
- Germany was represented by Dechema;
- Italy was represented by Italian Community of Adhesion.

Even though France is a prominent member country of EURADH, and the top five French universities were analysed, these French universities will not be considered in this study due to lack of results. Although several efforts were made to obtain information regarding the syllabus, researchers, French adhesion association and directors of mechanical engineering courses were contacted, however, no data were given on whether adhesive bonding was taught.

2.2 Selection of the Universities

The QS World University Rankings are annual university rankings published by Quacquarelli Symonds (QS), based in the UK. The top five Mechanical Engineering courses (according to QS World University Rankings, for 2020) in Germany, Italy, The Netherlands, Spain, Portugal and the UK were studied. Table 1 summarises the countries and universities selected.

2.3 Syllabus Contents Analysis

Following the QS Ranking data, the universities with the best ranked courses in mechanical engineering have been chosen. The syllabus contents of each select course was analyzed in order to understand in what context and how adhesive bonding technology was taught. In a first analysis, it was possible to identify that this subject

Table 1 Selection of the top 5 universities according to QS World University Rankings

Country	University ranking per country				
	1	2	3	4	5
Germany	RWTH Aachen University	Technical University of Munich	Universitat Stuttgart	KIT, Karlsruhe Institute of Technology	Technische Universitat Berlin
Italy	Politecnico di Milano	Politecnico de Torino	Sapienza University of Rome	Università di Padova	University of Naples—Federico II
Netherlands	Delft University of Technology	Eindhoven University of Technology	University of Twente	–	–
Portugal	University of Lisbon	University of Porto	University of Coimbra	Universidade Nova de Lisboa	Universidade of Minho
Spain	Universidade Politécnica de Madrid	Univesritat Politècnica de Catalunya—Barcelona Tech	Universitat Politécnica de Valencia	Universidade Carlos III de Madrid	Universitat de Barcelona
UK	University of Cambridge	Imperial College London	University of Oxford	The University of Manchester	Cranfield University

was taught at bachelor or master level. It should be noted that in the Portuguese system, until 2021, the bachelor's and master's degrees were merged into a single degree, called an integrated master's degree.

The syllabus contents were carefully analysed, understanding whether this technology was taught in just one or multiple subjects throughout the course, and to identify the didactic tools that were used to expose the adhesive bonding technology. In some cases, the didactic contents were not available to the general public for consultation. Thus, email contacts were held with course directors, chairpersons and researchers in the field of adhesive bonding, seeking to clarify whether and how this technology was being lectured.

3 Results

3.1 Ranking Position of the Selected Universities

Following the identification of the target universities, it was possible to understand the distribution of the universities and their position in the ranking. The QS rankings provide worldwide and regional tables (both at the country and continent level),

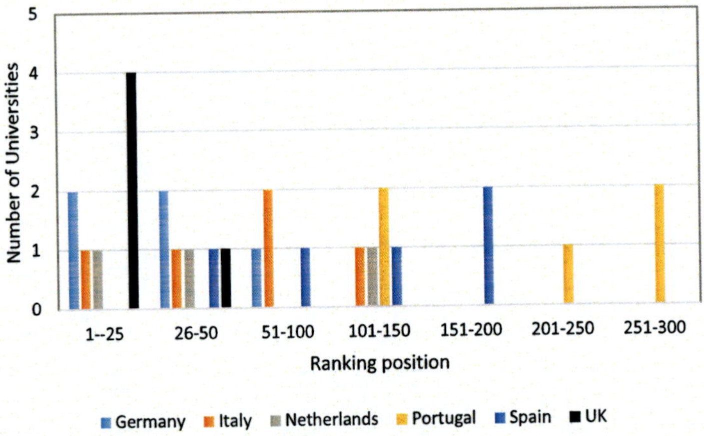

Fig. 1 Distribution of selected universities' according to their country and their ranking position

which are independent and include differences in the criteria and the relative factors used to generate them. The results of this ranking are presented in two different ways:

- Universities that rank up to position 50, are placed sequentially (i.e., 1, 2, 3 …). Information is available on their overall position.
- Universities after rank 51, are grouped in rank intervals (i.e., 51–100, 101–150, …). No information is available on their precise rank. The results of the overall score are determined by considering: academic reputation, employer reputation, faculty student ration, citation per faculty, international faculty ratio and international student's ratio.

Figure 1 illustrates how this distribution is carried out. It is observed that UK universities are those which are best ranked, featuring four universities in the world's top 25 and one university ranked in the top 26–50. Overall, universities from Central European countries (Germany, Italy and the Netherlands) are among the top 150 universities. On the other hand, universities from Western Europe (Iberian Peninsula) do have lower ranking positions. Portugal is the country with the lowest ranking universities; the five selected universities are ranked between the 101 and 300 positions.

Following this global analysis of the selected countries, a more detailed assessment was made considering the data available for each country.

3.1.1 Germany

The top five German universities selected are among the top 100 universities worldwide, with four of them being placed in the top 50. An analysis of the syllabus content of mechanical engineering courses shows that in 80% of the universities, the subject

Table 2 Top five mechanical engineering courses at German universities, considering their ranking position and whether they contain the topic of adhesive bonding in their syllabus

University	Ranking position	Overall Score	Syllabus content	Subject	Degree
RWTH Aachen University	18	86.9	Yes	2	BSc/MSc
Technical University of Munich	25	84.2	Yes	1	MSc
Universitat Stuttgart	42	80.3	Yes	2	BSc/MSc
KIT, Karlsruhe Institute of Technology	43	80.1	Yes	1	MSc
Technische Universitat Berlin	51–100	–	No	–	–

of adhesive bonding is taught at both the bachelor and master's level. These details can be found in Table 2.

3.1.2 Italy

As shown in Table 3, the Italian universities selected are among the top 150 universities in the world, two of them ranked above the top 50, two ranked between 51 and 100 and one university ranked among the top 150. Analysing the course contents, it is concluded that adhesive bonding is addressed in just 40% of the universities (at Politecnico di Milano only in Master's degree and at Universitá di Padova in Bachelor's degree). In 40% of the universities, it is not taught. Unfortunately, no data could be obtained for the University of Naples.

Table 3 Top five mechanical engineering courses at Italian universities, considering their ranking position and whether they contain the topic of adhesive bonding in their syllabus

University	Ranking position	Overall score	Syllabus content	Subject	Degree
Politecnico di Milano	9	90.4	Yes	1	MSc
Politecnico de Torino	30	82.9	No	–	–
Sapienza University of Rome	51–100	–	No	–	–
Università di Padova	51–100	–	Yes	1	BSc
University of Naples—Federico II	101–150	–	–	–	–

Table 4 Top three mechanical engineering courses at Dutch universities, considering their ranking position and whether they contain the topic of adhesive bonding in their syllabus

University	Ranking position	Overall score	Syllabus content	Subject	Degree
Delft University of Technology	5	92.9	No	–	–
Eindhoven University of Technology	37	81.1	Yes	1	MSc
University of Twente	101–150	–	No	–	–

3.1.3 Netherlands

Only three Dutch institutions were selected for this study since these are the only ranked universities teaching Mechanical Engineering. These Dutch universities are among the top 150 ranked universities. Delft University of Technology is ranked fifth worldwide, Eindhoven University of Technology is ranked at 37th and University of Twente in the 101–150 range. However, adhesive bonding is only lectured at Eindhoven University of Technology and in a master's degree, as shown in Table 4.

3.1.4 Portugal

Portuguese universities are among the 300 best universities in the world, as shown in Table 5. It can be observed that 80% of the selected universities teach adhesive bonding, in some cases in more than one subject, as is the case of the University of Lisbon and University of Porto, which shows that this is a topic seen by teaching staff of these universities as having significant importance. It should be noted the fact that this subject is taught in masters, but until 2021 the Portuguese educational system considered the integrated masters as a 5-year course, without providing the possibility of obtaining a standalone bachelor degree.

3.1.5 Spain

The selected Spanish universities are among the top 200 universities, as can be seen in Table 6. The Polytechnic University of Madrid, is the best ranked university, occupying position 49 in the world ranking. Adhesive bonding techniques are taught in 60% of the universities considered, always at Masters' level.

Analysing the geographical position of the universities, it can be observed that both universities in Barcelona do not include this technology in their study plans, unlike Madrid and Valencia Universities. The Universitat Politécnica de Valencia

Table 5 Top five mechanical engineering courses at Portuguese universities, considering their ranking position and whether they contain the topic of adhesive bonding in their syllabus

University	Ranking position	Overall score	Syllabus content	Subject	Degree
University of Lisbon	101–150	–	Yes	2	MSc
University of Porto	101–150	–	Yes	2	MSc
University of Coimbra	201–250	–	Yes	1	MSc
Universidade Nova de Lisboa	251–300	–	No	–	–
Universidade of Minho	251–300	–	Yes	1	MSc

Table 6 Top five mechanical engineering courses at Spanish universities, considering their ranking position and whether they contain the topic of adhesive bonding in their syllabus

University	Ranking position	Overall score	Syllabus content	Subject	Degree
Universidade Politécnica de Madrid	49	79.5	Yes	1	MSc
Universitat Politècnica de Catalunya—Barcelona Tech	51–100	–	No	–	–
Universitat Politécnica de Valencia	101–150	–	Yes	2	MSc
Universidade Carlos III de Madrid	151–200	–	Yes	1	MSc
Universitat de Barcelona	151–200	–	No	–	–

addresses adhesive joining in two subjects, which demonstrates the local academic community's interest in this particular topic.

3.1.6 United Kingdom

Of the countries selected for this study, the United Kingdom is the country with the best ranked universities, all of which are in the top 50. However, none of the selected universities teach this subject, neither at Bachelor's nor Master's level (see Table 7). Interviews were conducted with researchers (adhesive bonding research area) and mechanical engineering course directors at these universities to clarify this question, and they unanimously replied that joining technology is just taught and studied at the doctoral level.

Table 7 Top five mechanical engineering courses at British universities, considering their ranking position and whether they contain the topic of adhesive bonding in their syllabus

University	Ranking position	Overall score	Syllabus content	Subject	Degree
University of Cambridge	3	94.2	No	–	–
Imperial College London	7	90.9	No	–	–
University of Oxford	7	90.9	No	–	–
The University of Manchester	27	83.9	No	–	–
Cranfield University	43	80.1	No	–	–

3.2 *General Considerations on the Analysed Syllabuses*

An analysis was made of the syllabuses of mechanical engineering courses where adhesive joining is taught. Considering the syllabuses for 2020, some general conclusions and considerations can be made:

- Adhesive bonding is taught in a significant percentage of the mechanical engineering courses at the selected universities (43%). However, it is necessary to bear in mind that it was not possible to determine whether this technological process is taught (and how it is taught) for 18% of the selected universities.
- When taught, adhesive bonding usually is approached alongside other joining techniques or even unrelated technological processes, such as welding, riveting, casting, forming, among others. This technology is only taught separately at the doctoral level, which is entirely outside the scope of this study.
- In most of the analysed syllabuses, only a theoretical exposition of the didactic contents is used. This is a relatively new topic in most universities and thus a complete support structure (exercises, laboratory activities, etc.) for these classes is often not yet fully established. It is expected that as, the interest of the industry in this technology increases, other didactic tools will be included, providing a deeper teaching process.
- In contrast, this technology is already taught with a practical component in some institutions, such as the University of Porto. According to the interview with the director of the Mechanical Engineering course of this University, this technology is being increasingly applied in industrial settings and thus future mechanical engineers should understand its basic principles, process characteristics and understand its relative advantages and disadvantages, especially when compared with more traditional joining technologies, such as welding or riveting.

- In some universities, a dedicated educational trip has been conducted to consolidate knowledge and for students to experience industrial environments. This experience, which is considered to be extremely valuable, was highly compromised after the restrictions imposed by SARS-CoV-2.

3.3 *Knowledge Transfer*

To better understand if teaching adhesive bonding is correlated with the level of scientific research in this field, the scientific output of the same thirty-three Universities in the topic of adhesive bonds was also investigated. Data in the Scopus database was used for this study. Scopus is an Elsevier citation and abstract database, covering book series, journals and trade journals, having more than 41,000 indexed titles with a large temporal coverage (since 1788). To conduct this research, the authors delimited the keywords for "adhesive joint" between the years 2000 and 2020, for the countries considered in this study.

Generally, the authors observed that there has been a considerable increase the number of documents related to this technology published in the last two decades (120% increase), as shown in Fig. 2. When the document publication was analysed in a country by country basis, it was possible to see that the main drivers of scientific research in this area are Germany and the UK, as seen in Fig. 3.

It was also observed that all the universities included in this study do have publications dedicated to the topic of adhesive joining. Some of them are particularly active despite not including this joining technology in their study plans, as is the case of UK universities.

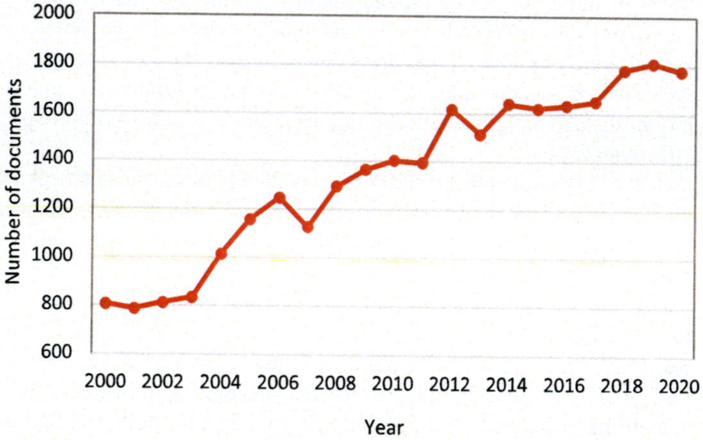

Fig. 2 Evolution of documents published from 2000 to 2020, in member countries of EURADH

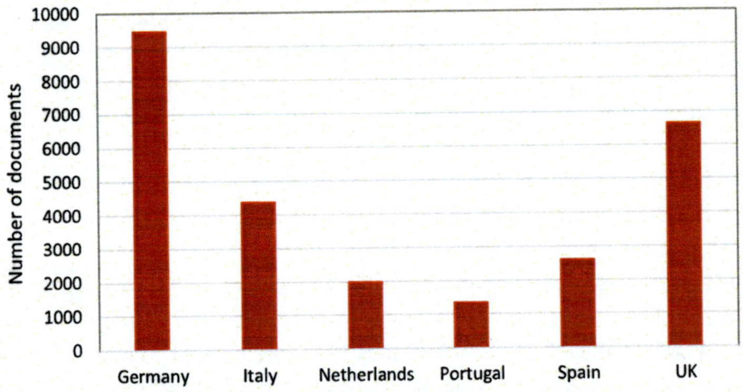

Fig. 3 Number of documents published by EURADH member countries, between 2000 and 2020

Although Portugal has a lower absolute contribution than the other countries analysed, it is the University of Porto which leads as the institution with the highest number of published documents (601) in Europe. This shows that this University and the research team integrated in it are highly committed to achieving technology transfer in this area of research.

3.4 *Specialized Training in Adhesive Bonding*

Another way of understanding how the scientific community is answering the training needs of the industrial community is through the analysis of specialised training programs. To quantify the courses on adhesive bonding available in each country, data provided by FEICA was analysed. FEICA is the Association of the European Adhesive & Sealant Industry. Created in order to support the industrial community in sustainable growth, promoting innovation, aiming to improve efficiency and effectiveness.

It should be noted that these courses generally possess a practical and less fundamental nature, especially when compared to what is taught in mechanical engineering courses. These courses are thus highly oriented towards the training needs of the bonders, adhesion specialists or even engineers who work with this technology on a daily basis. Figure 4 shows the courses taught in each country in 2020, according to information provided by FEICA. These courses are taught by universities and research centres dedicated to training. Figure 2 shows that Germany is the country that provides the most training in this area (19 courses), followed by the Netherlands. According to this data, Portugal and Spain are the European countries which provide the less significant training offer.

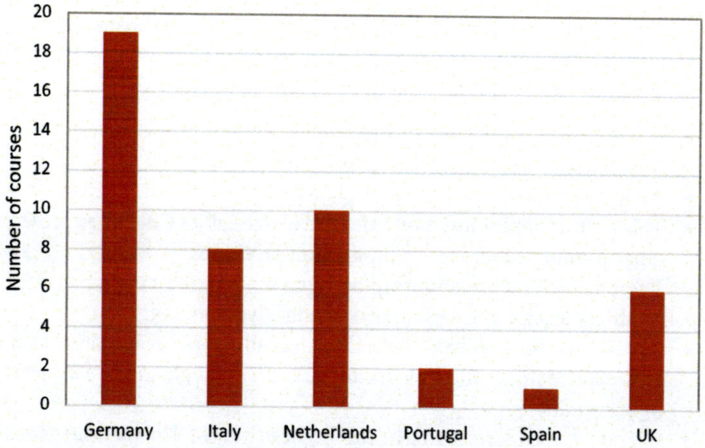

Fig. 4 Training offer in the EURADH member countries, according to FEICA

The large number of training courses offered in central European countries can be mainly attributed to the fact companies operating on the railway sector are required to have certified training for its workers.

The authors are aware that the information provided by FEICA may not be a factual reflection of the training reality in each country. It is possible that additional training might exist that is not available on FEICA's training dissemination channels. Nonetheless, this association does provide the mostly comprehensive listing of what training on Adhesive Bonding is available within Europe.

Comparing the results of Fig. 4 with those previously analysed in Fig. 1, it can be seen that there is no direct correspondence between the training offer and the existence of courses dedicated to adhesive bonding. Countries such as the UK, which do not have this technology in their degree programmes, do have a training offer. On the other hand, we observe that countries like Germany have already established the need to learn this technology, both at the university level and in technical courses. Although the authors were not able to obtain data for French universities, it can be observed that there is a large training offer in this technology. Portugal, on the other hand, has made a significant effort towards the training of new engineers with competences in adhesive bonding, while lagging in technician training processes.

4 Conclusions

An analysis was made of the curricula of highly ranked mechanical engineering teaching institutions of five European countries, seeking to determine if and how adhesive bonding is taught. A correlation between the amount published research in this field and the integration of adhesive bonding in the curricula was also made.

Lastly, professional training with these five countries was analysed in an attempt to understand the needs for qualified workers of the industrial sector. The following conclusions can be drawn:

- A significant percentage of mechanical engineering courses (43%) taught at the best European universities address adhesive bonding;
- Adhesive bonding is mostly taught at the Master course level;
- This topic is not exclusively dedicated to one subject, but often presented alongside other joining techniques or production processes;
- There is no direct relationship between the ranking position and the existence of a subject teaching topics related to adhesive bonding;
- UK has the best ranked universities, yet this topic is not taught on any of them. However, there are multiple technical courses taught outside the academic context of universities;
- Germany and Portugal have the highest rate of Universities teaching adhesive bonding;
- Portugal presents a high rate of adoption of adhesive bonding in its curricula, despite being the country with the worst ranked universities;
- It can be observed that the German education system is highly committed to this technology, since it is present in most of the mechanical engineering syllabus and also presents a high number of courses taught outside the university context;
- Although not all universities teach adhesive bonding topics, all have scientific research in the area. This shows that there is a transfer of knowledge and an awareness of the importance of this topic;
- Countries with the highest rankings are those with the most industry-oriented courses.

References

1. Kurfess, T.R.: Producing the modern. Engineer **19**(1), 118–123 (2003)
2. Borges, C.S. et al.: Women in mechanical engineering: a case study of the faculty of engineering of the University of Porto in the last 20 years **2**(1), 48–67 (2022). https://doi.org/10.24840/2795-4005_002.001_0005
3. Tryggvason, G., et al.: The new mechanical engineering curriculum at the University of Michigan **90**(3), 437–444 (2001). https://doi.org/10.1002/j.2168-9830.2001.tb00624.x
4. da Silva, L.F.M., Öchsner. A., Adams, R.D. (eds.): Introduction to adhesive bonding technology. In: Handbook of Adhesion Technology, pp. 1–7. Springer (2018)
5. Antelo, J., et al.: Replacing welding with adhesive bonding: an industrial case study **113**, 103064 (2022). https://doi.org/10.1016/j.ijadhadh.2021.103064
6. Silva, L.R.R., et al.: Polymer joining techniques state of the art review **65**(10), 2023–2045 (2021)
7. Marques, E.A.S., et al.: Introduction to Adhesive Bonding, p. 272. Wiley, Weinheim (2021)
8. Marques, E.A.S., et al.: Testing and simulation of mixed adhesive joints for aerospace applications **74**, 123–130 (2015). https://doi.org/10.1016/j.compositesb.2015.01.005
9. Borges, C.S., et al.: Influence of mode mixity and loading rate on the fracture behaviour of crash resistant adhesives **107**, 102508 (2020). https://doi.org/10.1016/j.tafmec.2020.102508

10. Delzendehrooy, F., et al.: A comprehensive review on structural joining techniques in the marine industry **289**, 115490 (2022). https://doi.org/10.1016/j.compstruct.2022.115490
11. Paiva, R.M., et al.: Adhesives in the footwear industry **230**(2), 357–374 (2016). https://doi.org/10.1177/1464420715602441
12. Cognard, P.: Handbook of adhesives and sealants: general knowledge, application of adhesives, new curing techniques. Elsevier (2006)
13. Barbosa, A.Q. et al.: AdTech project: European harmonized training system focus on adhesive bonding technologies **1**(1), 133–149 (2021). https://doi.org/10.24840/2795-4005_001.001_0006
14. Barbosa, A.Q. et al.: European adhesive bonder: a targeted training for Portuguese professionals harmonized with European directives **7**(1), 37–47 (2021). https://doi.org/10.24840/2183-6493_007.001_0006
15. Carew, A.L., Cooper, P.: Engineering curriculum review: processes, frameworks and tools. In: Proceedings of the Annual SEFI Conference (2008)
16. Carbas, R.J.C. et al.: Advanced joining processes unit: a fully independent research group **7**(1), 24–36 (2021). https://doi.org/10.24840/2183-6493_007.001_0005
17. Póvoa, L.M.C., et al.: Technology transfer from universities and public research institutes to firms in Brazil: what is transferred and how the transfer is carried out **37**(2), 147–159 (2010). https://doi.org/10.3152/030234210X496619
18. Dubickis, M. et al.: Perspectives on innovation and technology transfer **213**, 965–970 (2015). https://doi.org/10.1016/j.sbspro.2015.11.512

Learning Systems

Joint Designer: A Tool for Learning and Doing

Eduardo A. S. Marques, Ricardo J. C. Carbas, Marcelo Costa, and Lucas F. M. da Silva

Abstract Adhesive bonding is a nowadays a major joining technique for the manufacture of high-performance structures, seeing strong growth in diverse industries and applications, from the construction sector to the automotive industry. However, compared with classical joining techniques, it is still a relatively new process and one which still creates some challenges for the designers tasked with the engineering such connections. This difficulty is in part driven by the fact that teaching adhesive bonding is still very limited in most engineering courses, being taught without the depth that is necessary to train skilled designers, able to finely understand the specificalities of this technology. A solution for this issue relies in the development of educational tools that simplify the design process and accelerate the learning process. This work is related to such a tool, called JointDesigner, which has been the work of constant development in the Faculty of Engineering of the University of Porto and has been used both in an educational and industrial context to help design bonded joints. This work gathered feedback from the use of such tool and identified it as being effective in an educational context. Industrial users reported the tool suitability for the design of complex structures as limited but identifies the software as being highly appropriate for professional training and basic joint design purposes.

E. A. S. Marques (✉) · L. F. M. da Silva
Departamento de Engenharia Mecânica, Faculdade de Engenharia, Universidade Do Porto, Rua Dr. Roberto Frias, 4200-465 Porto, Portugal
e-mail: emarques@fe.up.pt

R. J. C. Carbas · M. Costa
Institute of Science and Innovation in Mechanical and Industrial Engineering (INEGI), Rua Dr. Roberto Frias, 4200-465 Porto, Portugal

L. F. M. da Silva and A. J. M. Ferreira (eds.), *3rd International Conference on Science and Technology Education 2022*, Proceedings in Engineering Mechanics,
https://doi.org/10.1007/978-3-031-25401-7_8

1 Introduction

Adhesive bonding is a fast-growing joining technology, which finds use in the manufacture processes of high-performance products, such as aircraft, road vehicles, ship building and in consumer electronics, among many other applications [1]. Although adhesive bonding has been used for many centuries, only during the last half of the twentieth century a significant effort was made to understand the mechanics behind these joints and thus lay the mathematical and physical groundwork necessary to design high-strength and durable joints, with the early work of Volkersen [2] and Hart-Smith [3], among many others. There is a significant existing scientific body of knowledge on how to precisely design bonded joints [4], but these procedures are often not taught to the students with the mechanical and aerospace engineering fields which will often later find themselves tasked with designing bonded joints later in their professional careers. In fact, there is a relatively strong training offer available for those are tasked to manufacture bonded joints (through on-the-job training and specialized courses) [5], but the design of bonded joints is rarely taught at an university level or, if taught, if often done in a passing or incomplete way.

Currently, the design of bonded joints is often done empirically and/or according to prescribed methodologies by engineers which do not possess formal training in this sort of activity. This is also compounded by the lack of certification procedures in this field (with the exception of the aerospace sector), which can make industrial users wary of investing in research and development activities, develop knowledge and adopt these technologies in its products. The largest and most technological driven of these companies address this issue via the implementation of complex finite element modelling procedures which necessitate the training of highly qualified personal and thus are only within the reach of a few select institutions. Nevertheless, even with these limitations the use of adhesive bonding is still growing. As a result, it has become highly important to provide simplified ways not only to help designers create strong and durable joints but also to understand the mechanics associated to this highly specific method of joining. This should be done in a context of academic education but also in industrial training settings.

As reported by Killen [6], any process of engineering education, independently of the field, is always a highly dynamic process, where pedagogical processes must be constantly evolved and contextualized to an ever-changing social context. The work of Reich et al. [7] reinforced this notion by identifying the fact that engineering education should always include a significant contribution from industrial requirements and formative processes, ensuring that the trained engineers are focused on practical industrial challenges.

One of the technological solutions which has been most extensively explored to support engineering practice and education have been Information and Communications Technologies (ICT) tools. With the development of the first electronic computers in the second half of the twentieth century, it eventually became possible to create tools that not only were used to design and engineering complex structures and systems but also simulate their behaviour in a manner that can be explored for education purposes in diverse fields, such as mechanics[8, 9]. These tools gained popularity as a simple and effective replacement for costly and time-consuming laboratorial activities [10].

A significant body of work has been published on the effect of using Information and Communications Technologies (ICT) tools in teaching technological subject matters. Still in the early stages of the development of the world-wide-web, Paterson et al. [11] presented a study on the environmental engineering student perceptions of the use of Internet-based learning tools. Students generally assessed web-based learning approaches as useful but were quick to stress that the use of conventional educational approaches are still viewed as important and cannot be fully replaced by online tools. The work of Abdulrasool et al. [12], carried out a study exploring the use of computer technology tools to improve the teaching–learning process in technical and vocational education settings. In their work, the group exposed to computer assisted instructions was found to perform much better than the group exposed to traditional teaching methodologies only. The work of Maragatharaj et al. [13] was devoted to the use of ICT tools to assist in the visualization of complex phenomena during teaching of complex phenomena, such as electromagnetic radiation emissions. Their work demonstrated the use of these tools to assist in the visualization was crucial to help students understand highly abstract concepts such as antenna radiation pattern and wave propagation. Santos et al. [14] carried out a work devoted to exploring the use of ICT, developing a tool called ISETL (Integrated System for Electronics Technology Learning) to facilitate the teaching process of electronic fundamentals. The use of this of this tool was found to significantly improve the students' learning outcomes.

Historically, a powerful design tool at the disposal of the designer of bonded joints are analytical models. At its most basic, analytical models are equations or series of equations which take into account the different geometrical characteristics of the joint and the mechanical properties of the adhesives and adherends used and return the stress state within the joint or the actual failure load of the joint. These models can range greatly in complexity, from very simple equations to complex. Ultimately, one of the most powerful methodologies to simplify the use of these models is to make of use of software tools that not only streamline the calculation process but, most importantly, can provide guidance with regards to the most suitable model for a given application. Through dynamic programming, it becomes possible to guide the user to a specific model based on, for example, the available entry data, the joint configuration or even the desired output.

The use of software-based design tools based on analytical models is of course not new. In fact, even for adhesive bonding there is a long history in this regard. The first computer programs used to design bonded joints was in fact proposed by Hart-Smith in 1973 [15]. This researcher was the first to provide a FORTRAN based tool for analysing different types of joints, including single and double lap joints, targeting aircraft construction applications. This tool was later part of the Primary Adhesively Bonded Structure Technology (PABST) program of the United States Air Force. The adopted models generally relied on simplified assumptions, since they did not account for the shear deformations within the adhesive layer and some of the analyses do not allow to determine the peel stresses, which are crucial for the design of composite bonded structures. From here onwards, other software-based solutions, has become available to achieve similar purposes. In 1992, Adams and Mallick [16], in the course of their research into the mechanics of bonded joints, proposed the JOINT software. The models include a formulation able to take into account the effects of tensile, shear and bending loads, as well as the possibility of including the effects of hydrothermal deformation in all components of a single lap joint (adherend and adhesive). Just a few years later, the University of Surrey developed a software known as the Stress Analysis for Adhesive Structures (SAAS) [17]. This software allows to create more complex joint models via a building block approach, in an effort to widen the applicability of the tool and used a very simplified finite element calculation method. An example of the SAAS interface is shown in Fig. 1.

More recently in 2003, Henkel AG and three Italian Universities developed a software known as JointCalc which, although targeted to expert users was still had some considerations to be intuitive for non-expert users [18]. This software included most joint configurations encountered in industrial practice. In addition to joint design, research applications have also benefited from the development of highly specific design software, such as modelling of the adhesive flow [19] and the accurate detection of the failure processes [20].

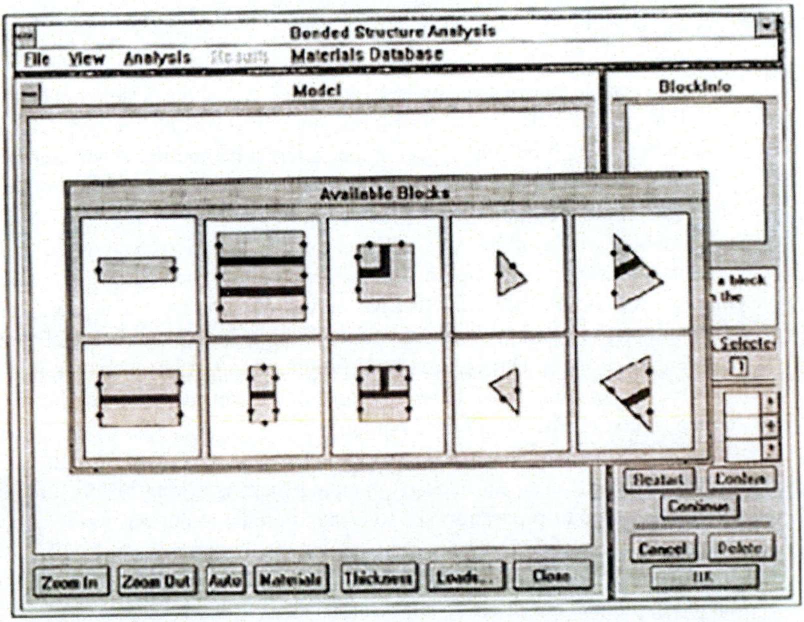

Fig. 1 Interface of the SAAS software [17]

Although powerful, most of these software solutions are still quite specialized and generally complex to use and hard to implement in training activities (be it in academic or professional settings). Furthermore, the resultant programs are also relatively hard to access and implement in practice, being mostly based on proprietary coded solutions. Perhaps due to these facts, no assessment of their usefulness has never been made or at least published to the best knowledge of the authors. Given this context, a novel adhesive joint design tool known as JointDesigner has been developed at the Faculty of Engineering of the University of Porto. This tool follows a web-based design, with a simple interface with guides the user in its operation. The aim of this study is to demonstrate how this custom designed joint design tool can be effectively used to improve the process of teaching the key design techniques of adhesively bonded joints in an academic environment. Furthermore, it is also intended to show that the same tool can be used to facilitate joint design procedures in an industrial environmental, helping designers to understand the key parameters which govern joint performance. The assessment is made through a set of questionnaires (for university students) and targeted interviews (for the industrial users).

2 The JointDesigner Software

As shown in the introduction, although diverse adhesive calculation software exists, they have been mainly designed with industrial or highly technical applications in mind. Furthermore, they all exist as standalone solutions with limited access. In this section, the software analysed in this work, JointDesigner, is presented in some detail, highlighting the aspects which facilitate its use in learning processes in industrial or academic settings.

2.1 Software Development

The first iteration of JointDesigner was programmed in MATLAB and was first published in 2009 [21]. It was designed to present a simple interface to the user, as shown in Fig. 2. Since it was programmed as standalone unit, this solution required a direct installation in the computer of the end-user.

This model was able to operate with single-lap, double lap and sandwich joints and could perform calculations according to the Adams et al. [22], Bigwood and Crocombe [23], Goland and Reissner [24], Hart-Smith [3], and Volkersen [2] analytical models.

Generally speaking, these analytical models provide direct equations that, when fed with the geometrical data of the joint and the material properties of the adhesive and materials being bonded (the adherends) will allow for a precise calculation of the level of stress within the joint. Thus, by ensuring that the stress level within the joint will not exceed the strength of the materials being used, one can design safe

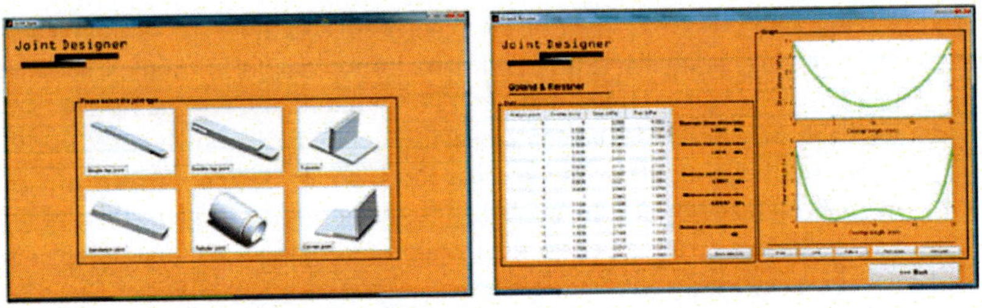

Fig. 2 First iteration of the JointDesigner software, proposed as a standalone software package

and durable bonded joints. Other specific features of this solutions were the ability to define how many data points the user would like to plot, a function to export the numerical data to other applications (by way of exporting to an excel file), and the ability to print resulting charts.

Eventually, the decision was made to transfer this application to a web interface, which allows the application to be accessed from any place or device with an internet connection. The tool was coded in PHP, which was deemed as a suitable general language and is supported by all web server hosting providers. The key advantages behind this approach were:

- Complete freedom to create the interface/application, limited only by the HTML code used to plot the software within the users' browser;
- Compatibility with any operating system, since any device at any place in the world can access the web application and use it without limitations provided, they have valid credentials;
- A large number of plugins/packages available that can provide assistance in the development of the product.

Of course, there also some disadvantages reported with this type of approach. The most important are:

- The requirement of using of a compatible Linux-based calculation solution to help with the more intensive calculations;
- The need to use a hosted server, which requires the payment of hosting and support fees that would not be applicable to a standalone application.

The interface of the second iteration of the JointDesigner tool is shown in Fig. 3 and is currently hosted on the www.jointdesigner.pt web address.

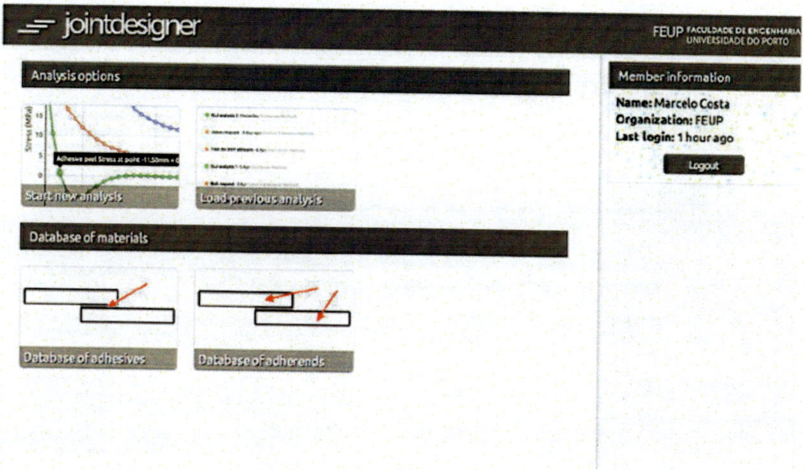

Fig. 3 Second iteration of the JointDesigner tool (as a web application)

2.2 *General Characteristics of the JointDesigner Software and its Operation*

Given its web application nature, the current version requires a user to create an account, which allows to save material data and previous analysis results. Upon successful login to JointDesigner, the user has multiple choices available, such as starting a new analysis, loading a previously saved analysis and consulting and altering the database of materials (both adhesives and adherends).

By selecting the first choice (new analysis), the user will be then prompted to select the joint geometry (single lap joint or balanced double lap joint are currently available) and then will be taken to a page where it can input the mechanical load acting upon the joint, the material properties (of the adhesive and the adherend) and the geometrical dimensions of the joint (Fig. 4).

Once the requested data is introduced, the software will automatically determine what analytical model can be used to calculate the stress present in the joint. A list of the models available in the software and its associated capabilities is shown in Table 1.

After selecting the desired analytical model, the software will run the necessary calculations and then output the predicted stress field within the joint. By adjusting the load applied to the joint, the user can seek a state where the stresses present within the adhesive do not exceed the yielding limits of both the adhesive and the adherends, thus determining a safe load that the joint can sustain. This process is depicted in Fig. 5.

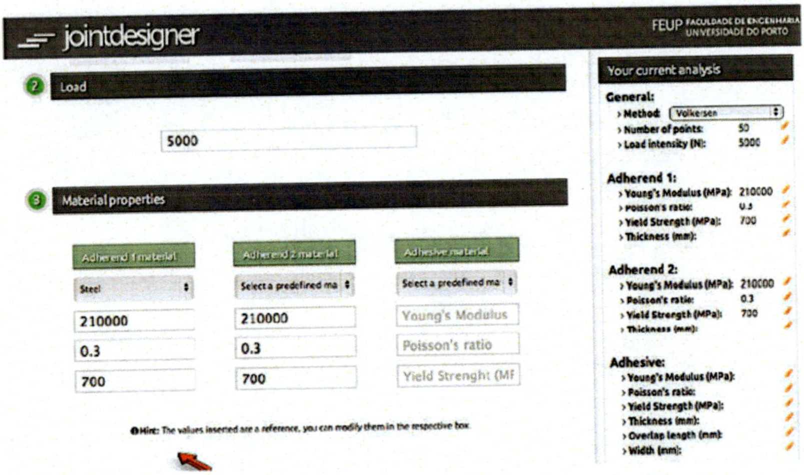

Fig. 4 Definition of the applied load and of the key material properties

Table 1 Analytical models supported by the JointDesigner software and its outputs

Model	Elastic adhesive	Plastic adhesive	Elastic adherends	Plastic adherends	Isotropic adherends	Outputs
Volkersen [2]	x		x		x	Shear stress
Goland and Reissner [24]	x		x		x	Shear and peel stresses
Ojalvo and Eidinoof [25]	x		x		x	Shear and peel stresses
Hart-Smith [3]	x	x	x		x	Adhesive shear, adhesive peel, plastic shear stresses and strains
Bigwood and Crocombe [23]	x				x	Shear and peel stresses
Adams [22]	x	x	x	x	x	Failure load

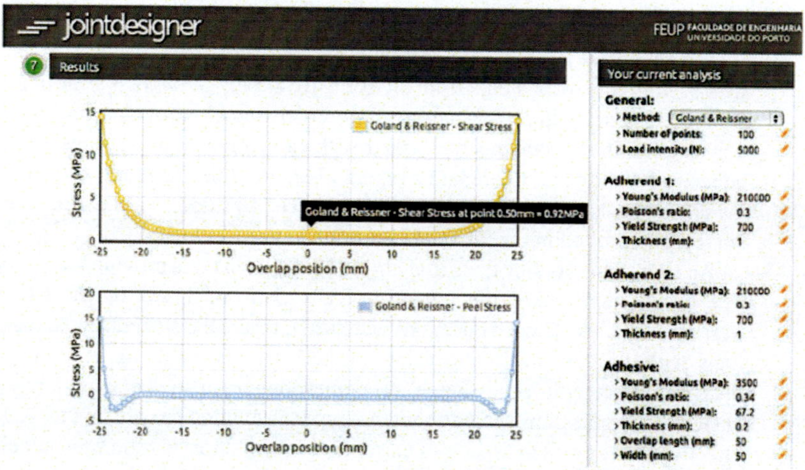

Fig. 5 Stress state within a single lap joint, calculated for a load of 5000 N

3 Research Methodology

3.1 *Generality*

As stated above, the main aim of this work is to determine if the JointDesigner tool can effectively flatten the learning curve associated to design of adhesively bonded connections. The methodology selected consisted in the realization of questionnaires and interviews of two different groups which can benefit from the use of this tool. The first of these two groups are students who use the software during their mechanical engineering degree and thus provide and academic perspective to the use of the software. Secondly, there is also the desire to gather an industrial perspective on the use of this software, achieved by communicating with design engineers operating in companies which use adhesive bonding in products.

3.2 *Data Collection*

3.2.1 Academic Feedback Collection

Modern mechanical engineering courses should be able to provide some level of training in adhesive joining in order to provide an answer for the industrial demands for this technology. The Department of Mechanical Engineering of FEUP offers three different levels of knowledge in adhesive joints to the students, spread over three different levels, with curricular units being proposed in the bachelor's degree, the master's degree and finally another in the Doctoral degree. The time devoted to the study of adhesive joining technique increases as a function of the level of study. The data used in this work stems from an extensive survey of the first semester students from the Bachelor in the "Manufacturing Processes I" curricular unit. During this curricular unit, the students are tasked with carrying out an experimental work on adhesive bonding and to prepare a report where the JointDesigner software is made available to calculate the joint strength. A total of 48 students has participated in the survey, all enrolled in the 2020/2021 academic year.

An anonymous online questionnaire of 12 questions was provided to the students. The questionnaire used the Inqueritos@UP platform, based on the LimeSurvey Version 3.28 code. Closed questions were used, where the students rated their agreement with different statements using a Likert scale (strongly agree, agree, neutral, disagree and completely disagree). The questionnaire also contained a few Yes/No answers. The questions versed about the student's opinion on software tools, learning effectiveness, as well as aspects of current implementation and some technical assessment questions. In the end, an open answer field was also included, prompting the student to freely identify some negative or positive aspects and provide suggestions of improvement.

3.2.2 Industrial Feedback Collection

A total of 12 different professionals from 12 different companies were interviewed. These companies were selected from different sectors, to ensure greater variety in the challenges and procedures used. The different sectors of which these companies belong to and the number of companies per sector are shown in Fig. 6.

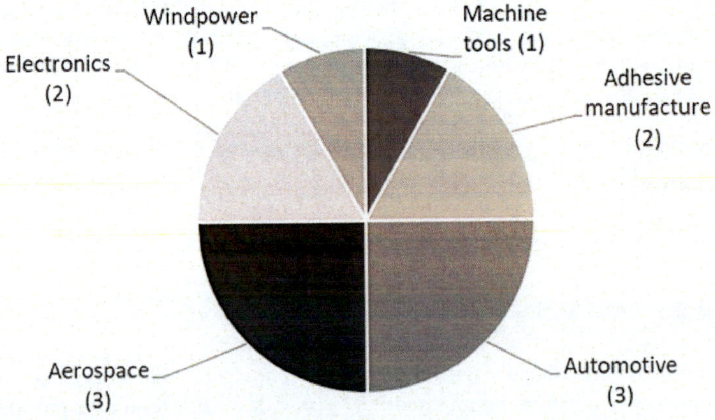

Fig. 6 Distribution of the industries (and users) targeted in the industrial feedback collection process. Number of companies per sector is shown in parenthesis

Four key topics were discussed with the professionals of these industries:

- Role of adhesive bonding in my company—understand how significant the usage of adhesive bonding within the interviewee's institution/company is.
- Tools for design of bonded joints with the interviewee's company—understand the tools which are currently used by the interviewee in its role as a designer of adhesively bonded joints and what needs does the company have in this regard.
- Training in adhesive bonding—determine if specific training is provided in the field of adhesive joining or if the interviews had to learn and develop procedures by themselves.
- Awareness and opinion of the JointDesigner software—this discussion topic had a two-fold aim. First, it was desired to understand the level of awareness to the existence of this software. Secondly, access to the software was provided to the interviewee, which tested the features of the software and provided an informed opinion on the suitability of this software for his/her intended purposes.

It is important to note that two of the interviewed companies have already integrated the JointDesigner tool in their design processes, albeit partially. These two companies operate in the automotive sector.

4 Feedback from Students

As referred above, student feedback was gathered through a questionnaire, designed to assess the prior knowledge of the student on the subject as well as its opinion on the software operation and the usefulness as complementary study tool.

4.1 Closed Questions

The analysis will be started by considering the answers to the questions posed to the students on their level of knowledge of the subject matter, the results of which are shown in Fig. 7.

Overall, the most significant conclusion is that the while the students report very limited or inexistant knowledge on the adhesive bonding subject, there a strong level of interest on this type of technology. Furthermore, students agree that they are highly likely to encounter adhesive bonding during their professional career. Also interestingly, almost a quarter of the students report that they already had designed some sort of structural joint, even if they did not have the necessary tools to do so.

Following this initial assessment, the next set of questions is concerned with operation of the software itself, as depicted in Fig. 8.

From this set of questions, one can easily verify that the students were generally pleased with the operation of the software, as most have reported a simple and clear operation. Furthermore, a large number of students also indicated that the information

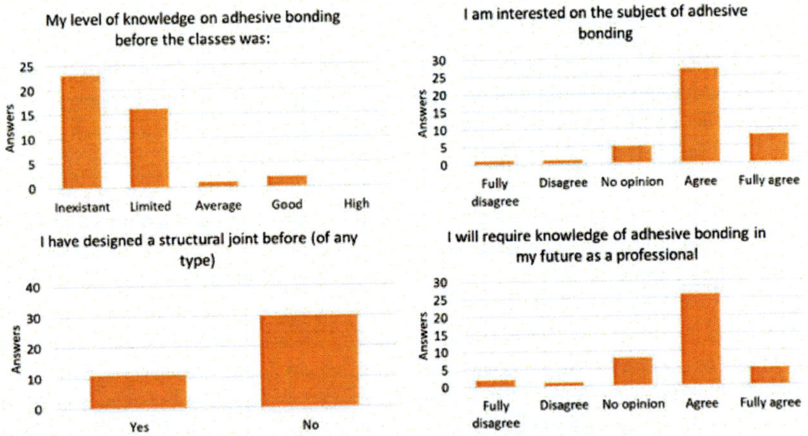

Fig. 7 Results of the questions related to the a priori knowledge of the subject matter

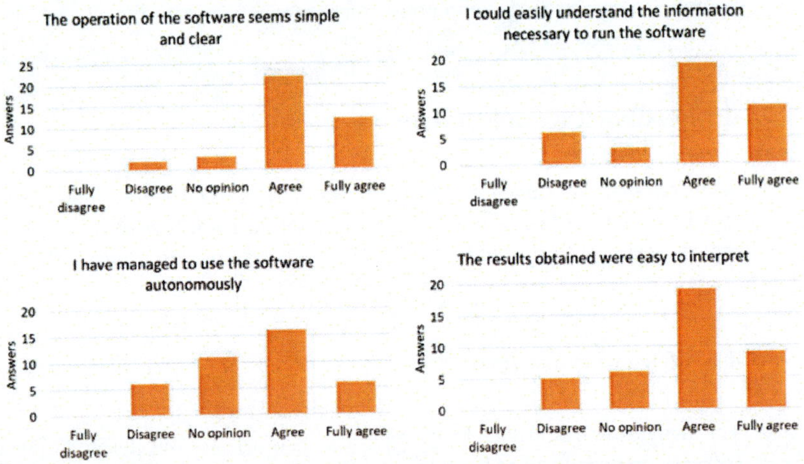

Fig. 8 Results of the questions related to the operation of the software

necessary to run the software was easily understandable, although a still significant number of students (around 10%) seemed to have some issues in this regard. The autonomous usage of the software was definitely a more contentious issue. Although the majority of the students still reported a successful autonomous usage, around 40% of the students seemed to have been challenged by this. This might be indicative of the fact that, although the calculation has been simplified by the software, the students feel they are still operating with a highly specialized subject matter, which requires, a least, a limited mastery of the underlying principles associated to join design work.

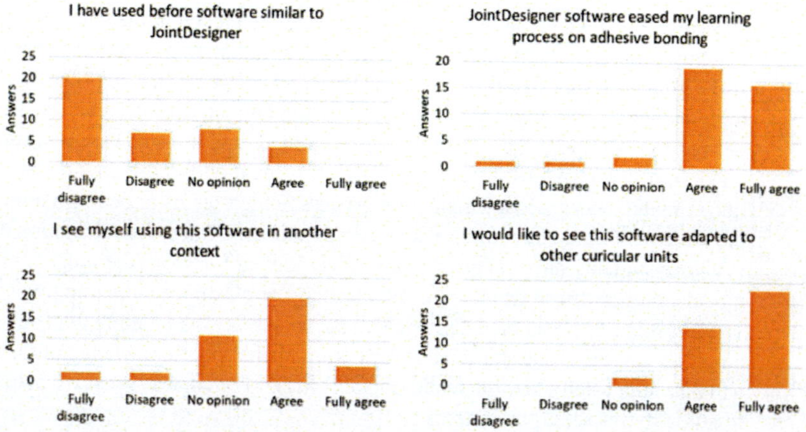

Fig. 9 Results of the questions related to the effectiveness of the software and potential future usage

A strong majority of the students reported that the results outputted by the software were easy to interpret. The percentage of those which did not agree or did not have a clear opinion was reduced to around 12%. This allows to reach the conclusion that while the operation of the software was challenging for some students, the same was less true for the interpretation of the results.

Finally, a last set of questions were placed regarding the effectiveness of the software as a part of the learning process and to determine if the students deemed it useful for future usage. These results are shown in Fig. 9.

With an analysis of the last set of datapoints, it becomes evident that the queried students do not have a lot of experience using software of this type, with only 10% of the students reporting having used before this type of software. Still, there was almost a unanimous opinion that this software as powerful tool for learning this technique. Furthermore, students appear to believe that they will quite likely use this software in another context (probably in a professional capacity) and that this type of software has a strong potential to be used in other curricular units.

4.2 Open Questions

In the open field students were asked to report what issues they face with the usage of the software and to provide some suggestions of improvement. The number of responses to these questions was much more limited but still represent valuable feedback.

A few students reported that the operation of the software could be improved by allowing for more modifications to be made to already calculated models, namely by allowing to change some of the input data which is locked in the calculation is run. Another common suggestion was to change the calculation process. Currently, the software requires the definition of a load level and will return the stress level. Accordingly, a couple of students have suggested the implementation of a reverse calculation whereupon the maximum stress level is identified, and the software will return the maximum load sustained by the joint. In fact, some models have already been modified to include such a load calculation process, which seems to be more intuitive for the student although it requires a more computationally intensive calculation process.

Other students have reported some difficulties in selecting the most appropriate models, namely those which consider elasticity or plasticity, something which has a very significant impact on the predicted load values. However, more than one student admitted that this issue stemmed mainly from his/her poor knowledge of the models and thus suggested the implementation of some sort of context sensitive help pop-up to guide in this selection process.

5 Feedback from Industrial Users

The interviews with engineers working with adhesive bonding in industrial settings generally reported a growing use of adhesive bonding in the industry, especially within the transportation sector. However, most of the interviewed reported that they received little or no training in the design and manufacture of bonded connections. Engineers tasked with producing the designs for bonded component and structures are mostly mechanical and material engineering majors and have also reported to have had limited formal training on this joining technique during their academic formation.

It was generally reported that the industry makes very limited use of analytical models. These models were initially developed for the aerospace industry and are mostly used at this level. The design approach in this industry is to follow standardized design procedures which seek to limit the stresses present within the joint.

In contrast, engineers working within the automotive sector do report a widespread use of commercial numerical packages (LsDyna for e.g.), which require complex material properties that can, in some applications, include the use of damage mechanics modelling. However, even those operating in this industry recognize their pre-existing limitations in the determination of necessary parameters to run these models and resort to training and material characterization activities provided by third parties, such as specialized research institutes. In fact, this is often regarded as the most challenging factor behind the numerical design of bonded joints and is commonly seen a risky proposition at the management level, since a significant investment must be made on this novel technological solution while the use of fasteners or welding is more mature.

Regarding the JointDesigner tool itself, multiple industrial users have described it as being quite useful for understanding the mechanics of a bonded joint. It's easy to use interface was praised, since it allows to quickly perform parametric studies and

Table 2 Most common positive and negative outlooks on the JointDesigner tool from an industrial perspective

Positive outlooks	Negative outlooks
Appropriate for initial training	Limited in joint geometry
Easy to use interface	Suitable only for initial joint design
Availability of pre-loaded material data	

thus determine the joint configurations which perform better under a given loading condition. In addition, the fact that some material characterization data is already preloaded in the software was also reported as being highly valuable, given the previously described issues with the adhesive characterization and the determination of reliable material property data. However, most users also noted that extended use of the application is mainly hampered by limited joint configuration geometries available, which mighty not follow the needs for complex joint geometries faced within the industry. This, of course, is driven by the fact that analytical models are quite complex to determine and thus mostly targeted at simpler joint geometries. Thus, there a general consensus that the use of the tool was more appropriate for initial training of professional and also a method for basic initial design of simpler joint geometries. In fact, the two companies which have partially implemented the joint designer tool in their practices do so in this manner. A summary of the most common positive and negative outlooks given by industrial users is provided in Table 2.

6 Discussion

The results of this work, gathered from two vastly different perspectives paint a clear picture regarding the effectiveness and usefulness of the developed software solution. Ultimately, all of the feedback gathered indicates a good level of satisfaction associated to the use of the software, especially for learning purposes. Although some students have reported some difficulties in the operation of the software, their opinions seem to validate the use of the software in teaching the subject of adhesive joint design and suggest that similar tools could be used for unrelated but complex engineering subject matters.

While most of the students report little a priori knowledge of the field, they were all generally able to carry out the requested calculations and came out of this process believing that this tool helped them to better understand the subject matter. Given these characteristics, the software can also be used to carry out specific academic research, in the context of master or doctoral thesis.

From the industrial study, the potential behind the use of the tool exists but it is more limited. In fact, industrial users are mostly challenged by the fact that existing analytical models rarely cover the joint designs which are present in modern bonded

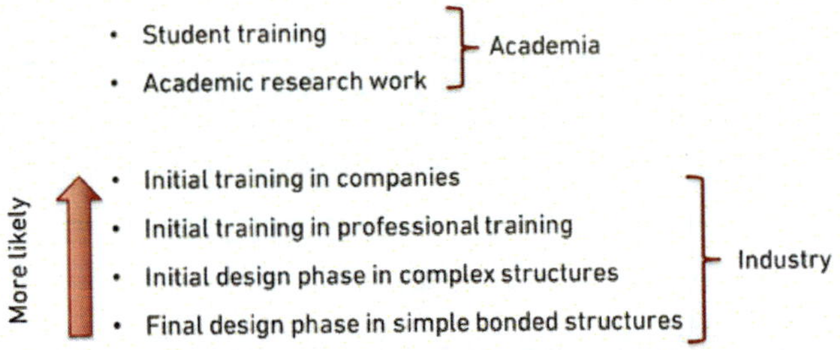

Fig. 10 Expected typical usage of the JointDesigner software in the industrial and academic sectors

products. Still, they recognize that the usage of most powerful design methodologies (such as finite element analysis) is often used without a detailed knowledge of the mechanics associated to the joint (a *black box* approach) and thus these software solutions can be effectively used to provide the designers with a simple method to understand how the performance of the bonded joint is affected by the different geometrical and material parameters. The software is more likely to be seen in the initial training phases of professions tasked with designing these joints and also deployed in a more limited role to create initial designs which might subsequentially be finely analysed through the use of more complex tools.

This analysis is summarized in Fig. 10, which provides the recommended and most likely usage of the software in the industrial and academic fields, considering the results gathered during this work.

7 Conclusions

This study demonstrated how a novel software tool can be used to positively reinforce the knowledge transfer within theoretical and practical classes in the field of adhesive bonding. Furthermore, it was also demonstrated this solution can be used to provide a support role in joint design work in an industrial setting, given the limited number of products available for accomplishing these tasks and the relatively low knowledge of the designers on the field of adhesive bonding.

The major conclusions of the work are:

- An online software tool was successfully developed to support teaching/training of adhesive joint design;
- Academic students reported little knowledge on adhesive bonding but were generally able to operate the software without major difficulties;
- The tool was reported easily accessible, simple to use and the results easy to interpret;
- Students report that this tool was a positive contribution to their learning process and the other unrelated subject matters could benefit from the of similar tools;

- Industrial users see value in using this tool for learning processes;
- Use in industrial design practices is restricted by the inherent limitation of analytical models.

References

1. da Silva, L.F.M., Öchsner, A., Adams, R.D. (eds.): Handbook of Adhesion Technology. Springer Science & Business Media, Berlin (2018)
2. Volkersen, O.: Luftfahrtforschung **15**, 41–47 (1938)
3. Hart-Smith, L.J.: Adhesive-Bonded Single-Lap Joints. Langley Research Center Hampton, VA (1973)
4. da Silva, L.F.M., Das Neves, P.J., Adams, R., Spelt, J.: Int. J. Adhes. Adhes. **29**, 319–330 (2009)
5. Barbosa, A., da Silva, L., Loureiro, A., Marques, E., Carbas, R., De Barros, S.: U.Porto J. Eng. **7**, 37–47 (2021)
6. Killen, C.P.: Eur. J. Eng. Educ. **40**, 476–498 (2015)
7. Reich, A., Rooney, D., Gardner, A., Willey, K., Boud, D., Fitzgerald, T.: Eur. J. Eng. Educ. **40**, 366–379 (2015)
8. Nikravesh, P.E.: Computer-aided analysis of mechanical systems. Prentice-Hall Inc. (1988)
9. Haug, E.J.: Computer aided kinematics and dynamics of mechanical systems. Vol. 1: Basic Methods. Allyn & Bacon Inc. (1989)
10. Yalcin-Celik, A., Kadayifci, H., Uner, S., Turan-Oluk, N.: Eur. J. Teach. Educ. **40**, 210–230 (2017)
11. Paterson, K.G.: J. Eng. Educ. **88**, 295–304 (1999)
12. Abdulrasool, S.M., Mishra, R.: Int. J. Learn. **15**, 155–168 (2009)
13. Maragatharaj, S., Rathinakumar, K.: Enhanced non-traditional learning environment for communication engineers using free OpenSource software tools. In: 2013 IEEE Fifth International Conference on Technology for Education (t4e 2013), pp. 101–104 (2013)
14. Santos, G., Mandado, E., Silva, R., Doiro, M.: Eur. J. Eng. Educ. **44**, 616–628 (2019)
15. Hart-Smith, L.: Adhesive-bonded scarf and stepped-lap joints. National Aeronautics and Space Administration, Contractor Report (CR), Jan 1, 1973 (1973)
16. Adams, R., Mallick, V.: J. Adhes. **38**, 199–217 (1992)
17. Gleich, D., Van Tooren, M., Beukers, A.: Progress in the development of an adhesive joint design and analysis tool. In: ICCM12 (2000)
18. Dragoni, E., Goglio, L., Kleiner, F.: Int. J. Adhes. Adhes. **30**, 267–280 (2010)
19. Aparecida Silva, L., Espinosa, C., Paroissien, E., Lachaud, F., da Silva, L.F.: J. Adhes. 1–33 (2021)
20. Ditz, M., Meschut, G., Teutenberg, D., Schwarze, T., Smart, D.: J. Adhes. **96**, 359–369 (2020)
21. da Silva, L.F.M., Lima, R.F.T., Teixeira, R.M.S.: J. Adhes. **85**, 889–918 (2009)
22. Adams, R.D., Comyn, J., Wake, W.C.: Structural Adhesive Joints in Engineering. Springer Science & Business Media, New York City (1997)
23. Bigwood, D.A., Crocombe, A.D.: Int. J. Adhes. Adhes. **10**, 31–41 (1990)
24. Goland, M., Reissner, E.: J. Appl. Mech. **11**, A17–A27 (1944)
25. Ojalvo, I., Eidinoff, H.: AIAA J. **16**, 204–211 (1978)

Assessment

The Importance of Laboratorial Classes Dedicated to Advanced Joining Processes in Undergraduate Engineering Education

Ricardo J. C. Carbas, Eduardo A. S. Marques, and Lucas F. M. da Silva

Abstract Laboratorial classes are an extremely important component of the learning process in undergraduate and post-graduate engineering courses. A well-designed laboratorial class often represents the best way to highlight and exhibit phenomena that are merely described in theoretical classes, capturing the student attention and boosting motivation for learning the topics approached in these courses. Laboratorial classes should target simple but effective case studies, aligned with the desired theoretical concept, seeking to provide necessary competences to students by requiring a good understanding of the underlying principles of the technologies involved and how these relate with practical applications. Accordingly, it is important to develop case studies that are as close as possible to actual practical industrial applications, facilitating the transfer of skills that allow for a better integration and understanding of current industrial needs. This work describes in detail a set of laboratory classes that has been developed in the Department of Mechanical Engineering of Faculty of Engineering of the University of Porto for curricular units associated with joining processes. The contents, objectives and the methodologies of the laboratorial activities are discussed. Furthermore, to evaluate the effectiveness of using these classes to further the learning process of the students, a query was prepared and distributed, allowing to reach the conclusion that the laboratory classes effectively aid the students and promote increased interest towards the class and its contents.

R. J. C. Carbas (✉) · E. A. S. Marques
Institute of Science and Innovation in Mechanical and Industrial Engineering (INEGI), Rua Dr. Roberto Frias, 4200-465 Porto, Portugal
e-mail: rcarbas@fe.up.pt

R. J. C. Carbas · L. F. M. da Silva
Departamento de Engenharia Mecânica, Faculdade de Engenharia, Universidade do Porto, Rua Dr. Roberto Frias, 4200-465 Porto, Portugal

L. F. M. da Silva and A. J. M. Ferreira (eds.), *3rd International Conference on Science and Technology Education 2022*, Proceedings in Engineering Mechanics,
https://doi.org/10.1007/978-3-031-25401-7_9

1 Introduction

Until the first half of the twentieth century, engineering education programs have mostly focused on preparing students by means of practice focused training. Science and mathematical models were used sparsely, and the necessary skills were developed mainly by imitation and repetition processes. However, engineering education throughout the twentieth century [1]. A greater scientific and mathematical modelling component was progressively included in education engineering programs, seeking to provide engineers with the necessary skills to develop complex and highly optimized projects, with higher performance and at lowered costs. Subsequently, the technological advancements of the twenty-first century led to accelerated changes within the engineering courses curricula, providing more research oriented skills and a greater emphasis on information technologies, reflecting the needs of a society which is now reliant on technological advances and growing digitalization [2]. Furthermore, in a twenty-first century context, engineering courses should graduate students which not only possess the necessary technical qualifications but also have the awareness and the drive to contribute towards a more efficient design and manufacturing processes, able to reduce the environmental footprint of the industries for which they will work and contribute to Bagherzadeh et al. [1].

This goal is only possible with the support of a faculty which possesses solid knowledge in fundamental concepts, carries out strong collaborative projects with the industry and fosters an intense and high-quality research activity. Nevertheless, effective teaching is only possible if the educator has true enthusiasm and passion for the subject matter and can transmit knowledge with clarity and in a well-structured manner [3–5]. One important tool to achieve this is the use of demonstrations, designed to clearly and easily illustrate the concepts discussed in class. In these demonstrations, simple objects and props can be used order to demonstrate basic concepts practice in a manner that can capture the attention of the students [6]. However, the implementation of these demonstrations within the classes should follow some basic rules. For example, there should be a careful selection of the most adequate demonstrations as a function of the concept being taught and set of research questions should be prepared to guide students towards a progressive and comprehensive understanding of the concept under study [7]. Ultimately, laboratorial classes should be designed to allow students to effectively and easily understand the theoretical concepts under analysis. In fact, laboratorial work can work as a powerful way in the process of acquiring new knowledge, from the evaluation and analysis of experimental results of particular phenomena or experiments [8, 9].

The aim of the current study was to highlight the importance of laboratory classes in an engineering education context and to describe methodologies for the organization of successful laboratory activities. In order to evaluate the effectiveness of the designed and implemented laboratory classes, a query was prepared, allowing the students to provide direct feedback on the impact of this type of classes in their learning process.

2 Methodology

2.1 Generality

Laboratorial classes can be considered as an effective complement to theoretical classes, furthering the understanding of theoretical concepts through simple experiments and/or demonstrations, highlighting fundamental concepts and allowing the students to manipulate specimens, demonstrators and technological equipment [10].

The main purpose of laboratory classes should be to motivate the students through intensive contact with the practical expression of the subject matter being taught. Here, this is achieved by introducing simple experimental works which ideally should be carried out by the students. Furthermore, when teaching technological processes, the use of modern equipment which is comparable with the technology in use industrial use is also essential to ensure an effective correspondence between what is taught and the challenges and tasks that the student will face during its professional career [11]. These types of classes, based on practical industrial case studies and problems, help the student to better understand the concepts and curricular units taught during the mechanical engineering courses [3, 12].

Moreover, it is crucial to ensure that the strategies and operational methodologies used for these laboratorial activities are allowed to evolve and not stagnate. Bear in mind that the educator has the full responsibility to be aware of novel pedagogical approaches or techniques to adopt in these works, which can better align with the students' needs and preferences and thus lead to more effective knowledge transfer [13, 14].

Moreover, laboratory classes should be set in environments with the necessary conditions for self-development of the students. This development is not only related to technical skills but is also extendable to a growth of the social skills associated to teamwork, such as conflict management, networking and communication. Laboratorial works often require the establishment of work groups where the students must work as a team and with a well-established leadership structure. While it would appear as ideal for a student to carry out experimental work individually, with his own laboratorial setup, this is often not possible, and the use of small groups can be used to retain the same level of knowledge transfer per student while adding a new dimension related to teamwork management.

The main technical and scientific skills developed by students include collecting, analysing, and interpreting data. Devoting our attention to the specific case of adhesive joining, students will contact and interact with diverse and unique equipment and techniques, something which is only possible in laboratorial classes and provides an opportunity to alert the students to the importance of safety procedures and equipment. In fact, laboratorial works can be considered as a sort of a first contact with a real-world scientific or industrial-like environment, one where unexpected situations are likely to occur and force the students to move away from their comfort zone.

In some cases, due to the complexity of the subject or the equipment, it might not be possible for the students to directly participate in the laboratorial work. An alternative might lie in the establishment of demonstrations, where a qualified technician will describe the key steps of the procedure and provide the necessary data for the student's report. [12, 15]

In addition to the development of teamwork skills, laboratorial work can also be used to promote a closer contact between the professor and student, which can have a positive effect in student engagement with the course and a desire to seek further information.

Laboratory work is typically concluded with a report, the preparation of which requires the student to exercise independent learning through the review of appropriate literature, manuals and the use of software tools, again providing the student with a another important skill for a successful engineering or science based career [16, 17].

2.2 Laboratorial Work Plan

All laboratory works or associated activities should have a well-designed plan, ensuring that the purpose of the activities are perfectly understood by the students [18, 19]. In order to help identify the main key points that a laboratorial work plan should follow, an example of an adhesive bonding laboratorial class will be first analysed in detail. This laboratorial plan was selected among the many that are currently used in the Mechanical Engineering course taught at the Faculty of Engineering of the University of Porto (FEUP) in Portugal, of which the authors are members.

The laboratory works or activities should have a well-structured plan in order to ensure that the procedure is perfectly understandable for all students. These laboratory plans need to include the purpose, background, work description, list of resources, experiment procedures and report guidance (Fig. 1).

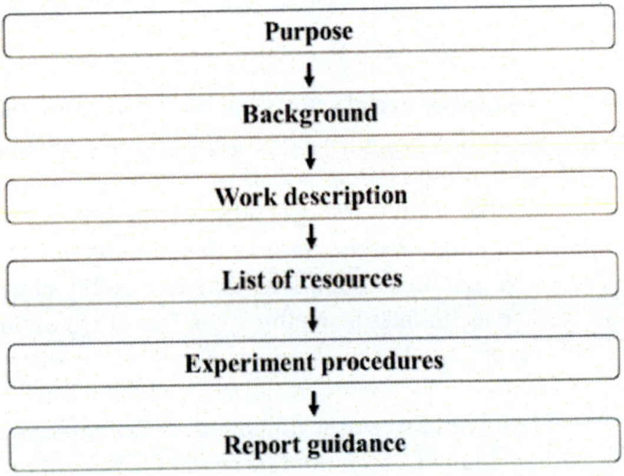

Fig. 1 Introduction section of laboratory plan dedicated to adhesive bonding

A laboratory work plan should clearly state the purpose, the background, all relevant theoretical concepts or principles, a complete work description, a list of resources, the detailed experimental procedure and strategy for analysing the results. The plan should feature a title that concisely describes the purpose of the experimental work to be carried out, include a brief introduction on the phenomena that will be observed and feature a description of all models or parameters that must be considered for proper understanding of the laboratorial work, addressing the theoretical part of the data being experimentally evaluated. Figure 2 shows examples of these documents, highlighting sections such as the title, the background and the theoretical background that supports the laboratory work. In this specific case, the effect of adhesive type on the strength of adhesive joints was evaluated.

Laboratory plans, such as the one shown in this example, must provide a work description with a simple explanation of the experiment design and a step-by-step guide for procedure that will be used. Whenever standardised procedures are to be carried out, the applicable standard should be included in the laboratory plan, allowing the students to analyse it and understand its essential characteristics. Moreover, it is necessary to clearly state which are the main results that should be extracted and what analysis procedures should be employed by the student/group.

The list of resources should define all the necessary materials, tools, instruments and any other equipment that should be used in the experimental work. It can aid the students in better understanding their work environment and become more familiar with the tools associated to a given process. Subsequently, a section detailing the experimental procedure is provided, whereupon all of the steps to be followed in the elaboration of experiment are described.

The last section of the laboratory plan should provide guidance for the preparation of a report. The nature of the report can, of course, vary, but in a general way, the students will be encouraged to provide a main result in clear manner and discuss these findings with the support of models and previous data. The students should not only be able to identify and describe significant trends in the results, but always be able to correlate this with the underlying mechanical, physical and chemical phenomena. According Bonwell and Eison [20], the preparation of a report is a simple but effective way to consolidate the concepts learned in the laboratory classes. It allows the students to reflect critically about the activities carried out during laboratory classes and how they relate with the different fields of knowledge that a mechanical engineering must master, or at least understand.

Figure 3 shows, examples of the report sections which are better aligned with the practical side of the activity. These include the work description, the list of materials, the experimental work description and the main discussion points that should be handled in the report.

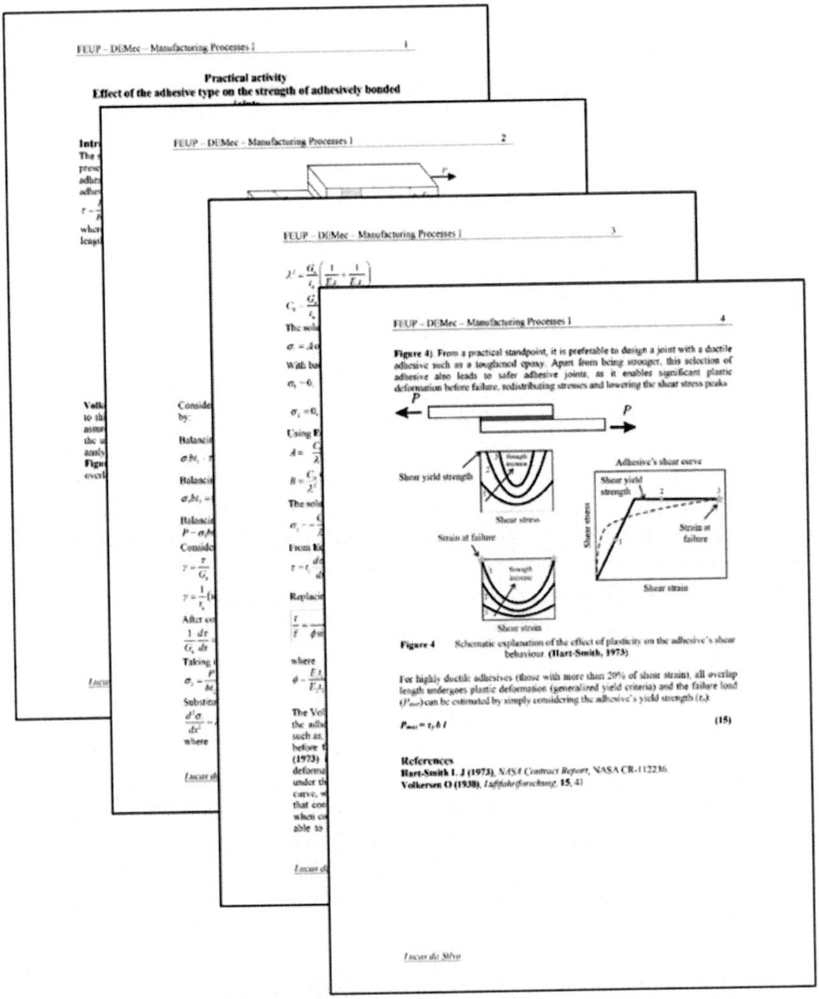
FEUP – DEMec – Manufacturing Processes I 1

Practical activity
Effect of the adhesive type on the strength of adhesively bonded joints

FEUP – DEMec – Manufacturing Processes I 2

FEUP – DEMec – Manufacturing Processes I 3

FEUP – DEMec – Manufacturing Processes I 4

Figure 4). From a practical standpoint, it is preferable to design a joint with a ductile adhesive such as a toughened epoxy. Apart from being stronger, this selection of adhesive also leads to safer adhesive joints, as it enables significant plastic deformation before failure, redistributing stresses and lowering the shear stress peaks

Figure 4 Schematic explanation of the effect of plasticity on the adhesive's shear behaviour (Hart-Smith, 1973)

For highly ductile adhesives (those with more than 20% of shear strain), all overlap length undergoes plastic deformation (generalized yield criteria) and the failure load (P_{max}) can be estimated by simply considering the adhesive's yield strength (τ_y):

$$P_{max} = \tau_y b l \quad (15)$$

References
Hart-Smith L J (1973), *NASA Contract Report*, NASA CR-112236.
Volkersen O (1938), *Luftfahrtforschung*, 15, 41

Lucas da Silva

Fig. 2 Example of a laboratory plan dedicated to adhesive bonding

2.3 Data Collection and Analysis

For this paper, two distinct sources of data were compiled and analysed. The data stems from an extensive survey of 1st-semester students from the Bachelor in the Manufacturing Processes I curricular unit, and also from the Advanced Joining Processes curricular unit lectured in the Master's degree. These units are lectured in the Mechanical Engineering department of the Faculty of Engineering of the University of Porto. A total of 68 students (45 Bachelor students and 23 Master students),

FEUP – DEMec – Manufacturing Processes I 5

Work description

Manufacture and determine the strength of:

- a single lap joint (SLJ) bonded with a ductile adhesive (MA422, Plexus);
- a SLJ bonded with a brittle adhesive (AV138/HV998, Huntsman);

The failure load of the bonded joint shall be determined using both the Volkersen model and the generalized yield of the adhesive criterion.

Materials

Substrates:

Table 1 High strength steel properties (DIN 55 Si 7 treated).

E (GPa)	ν	σ_y (MPa)
210	0.3	1100

Adhesives:

Table 2 Adhesive properties.

	MA422	AV138/HV998
Young's modulus, E (MPa)	994	4590
Poisson's ratio, υ	0.38	0.35
Shear yield strength, τ_y (MPa)	15	25
Maximum shear strain, γ_f (%)	>100	5

Experimental work

- Lightly abrade the bonded area with sandpaper.
- Clean the bonded area with a paper cloth embedded with acetone.
- Assemble the substrates in the mould and use spacers to obtain the intended adhesive thickness (0.2 mm) and overlap length (25 mm).
- Weight and mix the two components of AV138/H998 adhesive. Use 3.6g of AV138 (resin) and 1.4g of HV998 (hardener)
- Apply the adhesives on the bonded area.
- Join the substrates and close the mould.
- Apply the weights over the mould lid.
- Remove excess adhesive with a file and sandpaper.
- Mark and reference each specimen.
- Measure the final adhesive thickness using a calliper.
- Tensile test the complete joints in a universal testing machine, registering the failure load and the mode of failure.

Report

The report should include:

- prediction of the failure loads of the adhesive joints using the Volkersen model and the generalized yield of the adhesive criterion;
- plot with the failure loads and the failure load predictions;
- images of all the fracture surfaces and comments;
- comparison between predicted failure loads and experimental failure loads;
- indication of the better suited prediction method for each case under study and its justification.

Lucas da Silva

Fig. 3 Experimental information of laboratory work in adhesive joints have voluntarily participated in the survey, all students of the 2020/2021 academic year.

An anonymous online questionnaire, containing 19 questions, was provided to the students of both courses. The survey is mostly comprised of closed questions, where the students rated their level agreement with statements according to a Likert scale (strongly agree, agree, neutral, disagree and completely disagree). Furthermore, the survey also contained questions on the student's overall opinion of laboratory work, their impression on the learning effectiveness, as well as aspects of current implementation and technical assessment questions. In the end, a set of non-mandatory open answers were presented, providing a simple mechanism for the student to freely identify some negative or positive aspects of these laboratorial activities and provide suggestion of improvement.

3 Laboratorial Classes in Adhesive Bonding

3.1 Adhesive Bonding in Mechanical Engineering Course at FEUP

The use of adhesive joining technologies is steadily growing in many different industries and thus it is imperative to ensure that Mechanical Engineers finish their degrees with a good level of knowledge in this field, both theoretically and experimentally. Engineering courses should thus create the conditions for the design and implementation of experimental procedures and to provide the methods and knowledge for an adequate analysis of the problem at hand.

The Mechanical Engineering courses in the Department of Mechanical Engineering of FEUP offers the students three different levels of knowledge in the field of adhesive bonding. The first contact is made within the bachelor's degree. The subject can then be consolidated during the master's degree. Finally, the Doctoral degree courses allows for the student to carry out its own research, contacting with this field in a highly scientific manner. The time devoted to the study of adhesive joining technique increases with the level of degree, being higher for the doctoral degree.

3.1.1 Bachelor Degree—Manufacturing Processes I

In the Bachelor degree, the curricular unit that covers adhesive joints is Manufacturing Processes I. This is a mandatory unit that takes place in the third year of the bachelor degree. In this course unit, adhesive bonding is taught for one week, fully supported by theoretical and laboratorial classes. The time available for this curricular unit is a quite limited but it is still enough to instil in the students an important level of base knowledge of the adhesive joining technique. This is true even if only the effect of a limited number parameters on the performance of adhesive joints is explored and the laboratory classes are simple academic laboratory experiments, accompanied with simple analytical analysis to predict the mechanical behaviour of the adhesive joints.

3.1.2 Master Degree—Advanced Joining Processes

The Advanced Joining Process course unit is taught in the master's degree. This is an optional course and the time dedicated to adhesive joining technology is higher, taking place for 3 weeks and combining theoretical and laboratorial classes. In this unit, laboratorial works cover the effect of multiple different parameters on the performance of adhesive joints, such as the adhesive type, the adherend material, the effect of overlap length and the effect of the type of surface treatment. Furthermore, the laboratory work is always accompanied with joint design activities. In these activities, joint strength is calculated using analytical models and numerical analysis, with the obtained results being compared for accuracy against the experimentally obtained data. For this unit, an online analytical design tool (www.jointdesigner.pt) is used to help the students design joints and to understand how the different design parameters influence joint performance. The materials and surface treatments used are more complex [21, 22] and due this complexity the numerical analysis is usually conducted using Finite Element Analysis [23].

3.1.3 Doctoral Degree—Structural Adhesive Joints

The Structural Adhesive Joints course unit is also optional and offered for PhD students in their curricular year (in first year). In this unit, the laboratory works carried out are vastly different, deviating from the simple academic works used to evaluate the performance of adhesive joints. Here, students are integrated in research works and/or on going projects. This is possible because the student has to dedicate 3 months to this course, although in part time. Given these conditions, it becomes possible to integrate the students in research projects with a high level of novelty, allowing that the students develop strong experimental and numerical competences. The level of autonomy is of course much higher, as are the demands placed upon the student. Nonetheless, the opportunities provided for the development of personal skills, such as problem solving, team working, interpretation of results and scientific accuracy are much more significant.

3.2 *Proposed Laboratorial Works on Adhesive Bonding*

The laboratory works presented within this section seek to determine the influence of geometric parameters, the adherend type, adhesive properties and the surface treatments on the mechanical performance of adhesive joints [24]. These works have been developed for the mechanical engineering course in the Faculty of Engineering of the University of Porto. To evaluate the influence of use of laboratorial works on the students' understanding of adhesive bonding, a study was carried out by surveying students on the effectiveness and efficiency of laboratory works [25].

3.2.1 Effect of Surface Treatment

The first challenge associated to adhesive joints is the selection of an adequate surface preparation, a crucial step to ensure a good level of adhesion between the adherends and the adhesive. The literature proposes multiple solutions, ranging from a simple degreasing procedure with solvents, to a passive abrasive treatment to obtain a well-controlled level of roughness in the surface and the removal of a weak boundary layer. This can be obtained manually using emery paper or with the assistance of sandblasting machines. When the passive treatments do not ensure good adhesion, it is often necessary to use an active treatment such as anodizing for aluminium or titanium or the use of a plasma discharge for polymers.

In a laboratory work that studies the effect of surface treatment, different treatments are made available for the students to experiment with, and the effectiveness of surface treatment should be assessed through the determination of surface energy of adherends before and after the surface treatment. Finally, the strength and the failure mechanism of simple joints is evaluated through mechanical testing in a universal testing machine. In the end, the most desired failure mechanism is cohesive failure in the adherend, which means that the joint is not the weak point in a bonded structure. Also acceptable is cohesive failure in the adhesive layer, since the performance of the adhesive can be easily modelled. Adhesive failure, whereupon the adhesive is unable to adhere to the substrate is not acceptable.

3.2.2 Effect of Overlap Length as a Function of Adhesive Type

One other important laboratory work seeks to understand the effect of overlap length as a function of the type of adhesive that is used. In order to better analyse the effect of adhesive type and overlap length on the strength of the joint high strength steel adherends should be used, ensuring that the adherend remain in elastic domain during the test.

When a brittle adhesive (adhesive with no plastic deformation) is used in a joint, the joint strength is often limited by the strength of the adhesive used and the size of the bonded area increases (with the overlap length increase) there is a linear increase of the joint strength. However, this increase only occurs until a certain length is reached, as a plateau is eventually reached. This behaviour is due to the presence of stress peaks at the ends of the overlap. When the joint is gradually loaded, the stress peaks will eventually reach the ultimate strength of the brittle adhesive and the joint fails instantaneously. For this type of adhesive, increasing the overlap often does not represent significant increase in joint strength, which will remain almost constant.

In the other hand, when a ductile adhesive (with a shear elongation higher than 20%) is used in a joint, yielding of the adhesive will occur long before the joint fails. Such adhesives allow for an uniform redistribution of the stresses along all of the overlap length. This fact leads to a failure load that rises almost in a direct proportion to the overlap length.

3.2.3 Effect of Overlap Length as a Function of Adherend Type

The effect of overlap length as a function of adherend type should also be evaluated in a laboratory work.

When adherends with high strength are used and bonded with a ductile adhesive, one can find a linear relationship between the strength and the overlap length as was explained in the previous laboratory paragraphs. However, an alternative analysis is to consider joints using an adherend with low tensile strength a compare its performance with that of joints using high strength adherends, bonded with the same adhesive (ductile adhesive). For the low strength adherend, yielding will occur during the test, generating a high level of stresses on the adhesive at the ends of overlap length. This is mostly due to the rotation of the adherend, which promotes significant peel loads on the adhesive, causing premature failure. An increase of the overlap length does not promote a significant improvement of joint strength, as a plateau will also eventually be reached, even with a ductile adhesive.

3.2.4 Effect of Composite Adhesive Joints as a Function of Overlap Length

In order to complement the study of effect of adherend type, another laboratorial work should be devoted to studying the performance of adhesive joints when composite adherends are used. The peel strength (strength in the through the thickness direction) of composite adherends is generally low and this can lead to delamination failure when used in a joint. The failure mode of composites is also very dependent of overlap length and on the type of adhesive used. For joints with shorter overlap lengths, the peel stresses generated during the test at the ends of the overlap are not sufficient to promote delamination of the composite adherends. However, with an increase of the overlap length, the level of peel stress at the ends of overlap length is higher and this can cause a delamination of the composite adherend. Delamination is more likely should a brittle adhesive be used, as the stresses generated as more severe.

4 Results and Discussion

4.1 Closed Questions

Both groups of the Mechanical Engineering students (Bachelor and Master Courses) were assessed in what regards the general effectiveness of the laboratorial classes. The results of this analysis are shown in Fig. 4.

These two groups of students considered that these types of activity are important for complete understanding of theoretical concepts. Regarding manufacture processes, the laboratory classes were found to be effective, with the students reporting a complete understanding of all steps associated to the adhesive bonding process. For both groups, the overall assessment was almost the same, with the students reporting that the laboratorial classes were in fact important for them study to understand the concepts being discussed.

The student's assessment of the content of laboratorial classes held for each course is presented in Fig. 5.

Due to limited time available for adhesive bonding learning in the Bachelor course, it was only possible to introduce a single laboratorial work. For the academic year that was targeted with this query, the laboratorial work performed in the Bachelor course was concerned with the study of the influence of the surface treatment on joint strength. In contrast, the longer time available during the Master affords the development of different laboratorial works, covering not only the study of surface treatments but also the effect of type of adhesive and adhesive. It can be easily seen that, for both courses, students widely considered that the study of surface treatments in the joint strength of adhesive bonding is very important and that this was successfully accomplished by the students in the course of their participation in laboratorial works. Figure 5 shows that the main difference between both courses is

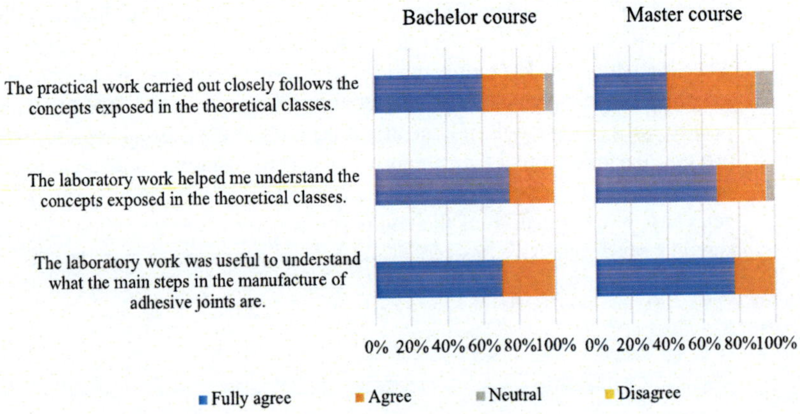

Fig. 4 Overall assessment of the laboratory classes

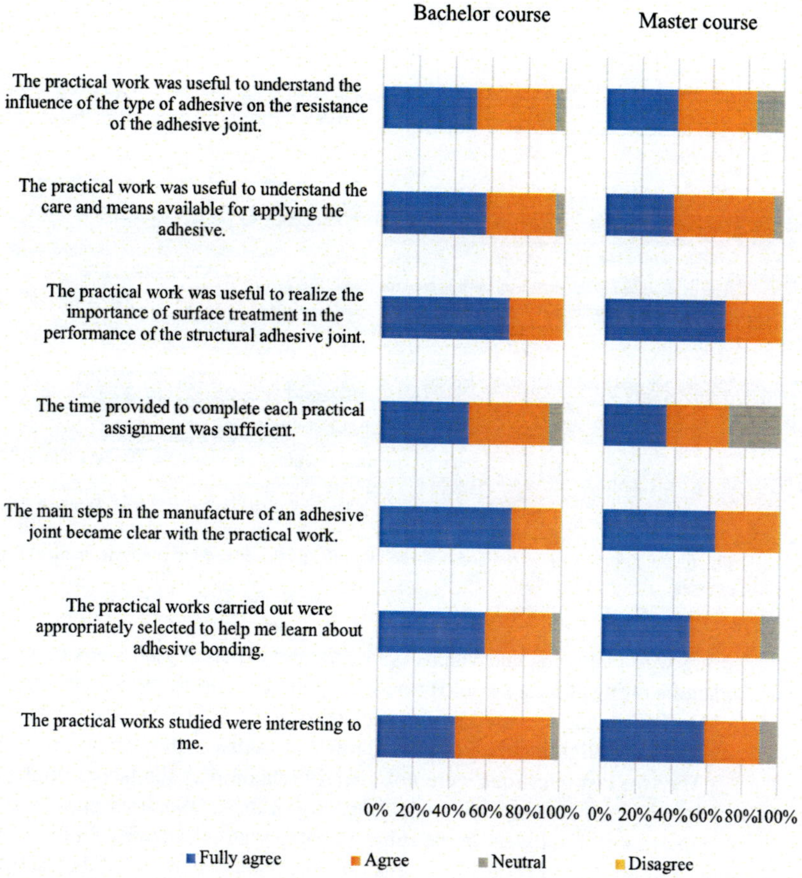

Fig. 5 Contents the practical works used in the laboratorial classes

mostly related to the time available to complete the laboratorial work. While students of the Master course are consistently satisfied with the time available, one third of the Bachelor students report the time available as short. Master students also considered that the practical works are generally more important than the Bachelor students, probably due to high level of demand placed upon the topic.

The pedagogical strategies used in the laboratory classes were generally considered as adequate, as can be seen in the Fig. 6.

The documentation and tools available to perform the laboratory work were identified as the key point that requires improvement, especially so for master courses. In Fig. 7, one can observe that over 50% of the master students considered that documentation being provided was not enough to carry out the laboratorial work and thus that the aid of a teacher or assistant was crucial to ensure that the correct procedure was adhered to. It is the authors opinion that additional support provided

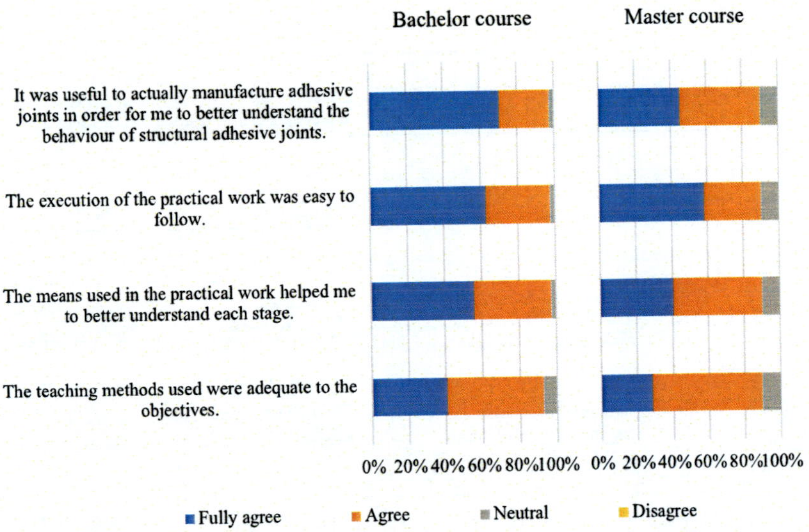

Fig. 6 Pedagogic strategies used in these laboratory classes used in both Mechanical Engineering courses

using web platforms and videos (showing and explaining all steps required in the manufacturing process) would provide a significant improvement in this regard. Such platforms can also be used by the student to prepare the laboratorial classes before the class, highlighting the key procedures that student will later have to follow.

The two last questions were not directly related with the laboratorial works but indirectly attempted to evaluate the capacity of these laboratory classes to improve the interest of the student in the subject treated with this activity. In Fig. 8, it can be seen that all students considered that the information learned will be useful in them professional life. Furthermore, a significant result obtained was the fact of almost

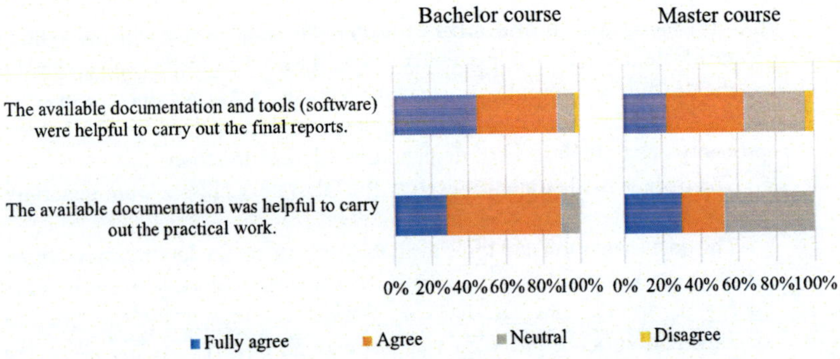

Fig. 7 Documentation and tools available for the execution of laboratorial work

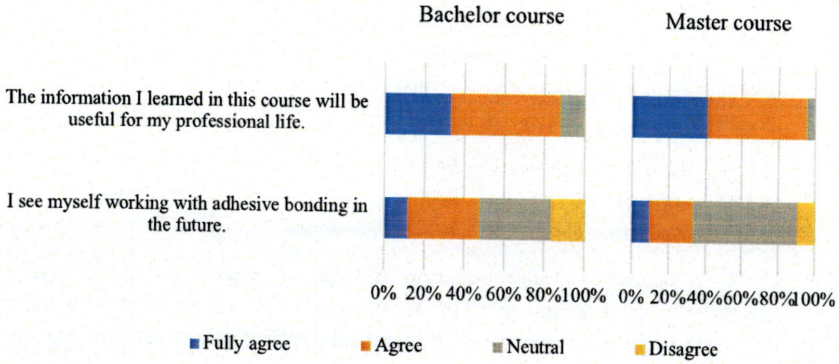

Fig. 8 Future perspectives of the students

50% of Bachelor students developed interest in working with adhesive bonding in the future. However, more than 50% of master students do not know if they would like to work with adhesive bonding. This could be explained by the fact by covering in deeper detail several aspects of this technology, so students are thus more informed of the nature of the adhesive bonding process and thus leads to a more clearly defined opinion in this subject.

4.2 Open Answer

In the open answer portion of the questionnaire, students were encouraged to identify a few positive and negative aspects of the laboratory classes and also to suggest some improvements that could be made to these classes.

The main positive aspects being indicated are:

- "*Very useful to relate the theoretical concepts with practice*"
- "*It is very motivating to have the chance to do things experimentally*"
- "*It should be expanded for other curricular units*".

In general, engineering students appreciate the experimental work and feel that this is key tool to correlate the theoretical concepts with practice. They believe that this way of learning is more efficient and that it should be extended to different curricular units.

The students also identified some important negative aspects:

- "*Time available for each class is short*"
- "*The number of laboratorial works is very limited*"
- "*Documentation could be more detailed*".

According to the first two points, the student appreciated the laboratorial classes but considered that the number of classes and its duration should be higher. This is, of course, in accordance with the positive points identified. The final negative point, regarding the level of documentation was already identified and discussed above in the section devoted to the closed questions of the quiz.

The main suggestions provided were:

- "*Recording the classes would help*"
- "*The effectiveness of could be improved with a bit more of time*"
- "*Increase the time supporting software tools*".

Class recording (and publication of these recordings) was a technique that was intensively used during the COVID pandemic. However, even with the return to presential modes of operation, students still regularly request this type of support tool, since it provides a highly practical way to recap the classes with precision and on-demand, aiding study. The duration of laboratory classes again was considered as limited, and the students considered that the effectiveness of laboratorial classes can be improved by increasing the duration of the activities. However, this of course requires changes to course planning and clashes with the need to allocate equal time for all of the course curricular units. One potential aid for this issue lies in the use of the support tools mentioned previously, which mighty allow the student to come well prepared for the laboratorial class and thus make better use of the short time available. Finally, the last suggestion is more adequate for the master course level, as the software used to support the numerical analysis is quite complex and thus requires significant training before the students can use the design software independently and competently.

5 Conclusions

This study presents the main advantages associated with incorporating laboratorial classes in engineering courses. It shown that all laboratory work should be accompanied by a well-structured work plan, allowing to ensure an effective understanding of the phenomena under analysis and its correlation with the underlying physical and chemical processes. In this study multiple works on adhesive joints have been discussed, which in general evaluate the effect of a set of key parameters on the mechanical behaviour of adhesive joints. Also listed are the essential steps that must implemented for the creation of a successful experimental work on adhesive joints.

The main points that can be considered for an efficient laboratory class can be listed as:

- An efficient laboratory work should be based on a good organisation of the workflow;
- In order to ensure that the laboratory work can capture the attention of the students, the work methodologies and strategies should be regularly updated. Furthermore, this drive for consistent updates allows to correct mistakes and avoid unwarranted difficulties felt by the students during the execution of the work;

- The student develops multiple skills by participating in these procedures. From soft skill, such as team building and management, to highly technical skills, such as the operation of complex equipment and modern manufacturing techniques, all of which with great relevance of the students future academic career;
- Ultimately, the goal should always be to ensure that the student concludes the work with aa simple and clear understanding of the theoretical principle behind each experimental work;
- By gathering the student feedback through a questionnaire type of tool it was possible to ascertain that;
- The students considered that the laboratorial classes are a useful methodology to improve knowledge retention and accelerate learning;
- Moreover, students considered that these types of classes should be implemented in other curricular units present within engineering courses;
- The time available for each laboratory work must be carefully monitored in order to ensure that all students completely finish the activity in order to optimize the effectiveness of knowledge transfer.

References

1. Bagherzadeh, Z., Keshtiaray, N., Assareh, A.: Eurasia J. Math. Sci. Technol. Educ. **13**, 6749–6760 (2017)
2. Grimson, J.: Eur. J. Eng. Educ. **27**, 31–37 (2002)
3. Singer, S.R., Hilton, M.L., Schweingruber, H.A., Us, N.R.C.: America's lab report: investigations in high school science. National Academies Press Washington, DC (2005)
4. Yalcin-Celik, A., Kadayifci, H., Uner, S., Turan-Oluk, N.: Eur. J. Teach. Educ. **40**, 210–230 (2017)
5. Khanna, S., Jenkins, C., Roylance, D.: Proceedings of the institution of mechanical engineers. Part L J. Mater. Des. Appl. **216**, 49–53 (2002)
6. Feisel, L.D., Rosa, A.J.: J. Eng. Educ. **94**, 121–130 (2005)
7. Brüggemann, J., Bizer, K.: J. Innov. Entrepreneurship **5**, 1–13 (2016)
8. Domijan, A., Embriz-Santander, E.: IEEE Trans. Power Syst. **7**, 1571–1578 (1992)
9. Eekels, J., Roozenburg, N.F.: Des. Stud. **12**, 197–203 (1991)
10. Hake, R.R.: Phys. Teach. **30**, 546–552 (1992)
11. Krivickas, R.V., Krivickas, J., Global, J.: Eng. Educ **11**, 191–196 (2007)
12. Newble, D., Cannon, R.: Handbook for Teachers in Universities and Colleges. Routledge (2013)
13. Dunne, J., Ryan, B.: (2010)
14. Kapilan, N., Vidhya, P., Gao, X.-Z.: High. Educ. Future **8**, 31–46 (2021)
15. Easton, W.: Education+Training (1959)
16. Council, N. R.: Science Teaching Reconsidered: A Handbook. National Academies Press (1997)
17. Hofstein, A., Lunetta, V.N.: Rev. Educ. Res. **52**, 201–217 (1982)
18. Wheeler, A.J., Ganji, A.R., Krishnan, V.V., Thurow, B.S.: Introduction to Engineering Experimentation. Pearson (2010)
19. Holman, J.: Experimental Methods for Engineers, McGraw-Hill Series in Mechanical Engineering. McGraw-Hill, New York (1994)
20. Bonwell, C.C., Eison, J.A.: Active learning: creating excitement in the classroom. In: 1991 ASHE-ERIC Higher Education Reports (ERIC) (1991)

21. Naat, N., Boutar, Y., Naïmi, S., Mezlini, S., Da Silva, L.F.M.: J. Adhes. 1–93 (2021)
22. Budzik, M.K., Wolfahrt, M., Reis, P., Kozłowski, M., Sena-Cruz, J., Papadakis, L., et al.: J. Adhes. **98**, 2133–2209 (2022)
23. Tserpes, K., Barroso-Caro, A., Carraro, P.A., Beber, V.C., Floros, I., Gamon, W., et al.: J. Adhes. **98**, 1855–1915 (2022)
24. Da Silva, L.F.M., Öchsner, A., Adams, R.D. (eds.): Handbook of Adhesion Technology. Springer Science & Business Media, Berlin (2018)
25. Marques, E.A.S., Akhavan-Safar, A., Tenreiro, A.F.G., Da Silva, L.F.M., Carbas, R.J.C.: Structural Adhesive Bonding in Aerospace Applications. Engebook, Porto (2022)

A Multi-dimensional Quality Assessment Instrument for Engineering Education

Leonhard E. Bernold and Brayan Díaz-Michell

Abstract The results of teaching evaluations are very often used by universities as a basis for personnel decision making, including promotion and tenure of instructors. The evaluations are commonly based on end-of-semester students' ratings of their teachers through the qualitative Likert-type Student Evaluations of Teaching (SETs). Unfortunately, research on the validity of the SET-type evaluations frequently produces misleading results. One serious problem is the use of survey questions that are qualitative rather than quantitative, result in evaluations that reflect the students' personal opinion of the class, rather than the effectiveness of the teaching. The focus of this paper is to present a design for an effective student evaluation that embraces the constructivist approach to teaching. The role of the teacher becomes that of a coach who creates a collaborative work environment, presenting relevant and interesting problems for students to solve. Assessing the educational quality of such a method requires a multi-dimensional instrument that qualitatively measures the effectiveness of both teaching and learning. Our study accentuates the need for an effective, knowledgeable teacher-coach, along with skilled, motivated students, and a supportive cohesive university atmosphere.

Keywords Teaching evaluation · Qualitative assessment · Constructivism · Multi-dimensional survey · Teacher efficacy

1 Introduction

The global need for more science, technology, engineering, and mathematics (STEM) graduates remains unfilled (Hossain and Robinson 2012; Olson and Riordan 2012). In order to meet the demands of the twenty-first century, many governmental initiatives

L. E. Bernold (✉)
Technical University Federico Santa María, Valparaíso, Chile
e-mail: Leonhard.bernold@gmail.com

B. Díaz-Michell (✉)
Department of Science, Technology, Engineering, and Mathematics Education, North Carolina State University, Raleigh, NC, USA
e-mail: badiaz@ncsu.edu

L. F. M. da Silva and A. J. M. Ferreira (eds.), *3rd International Conference on Science and Technology Education 2022*, Proceedings in Engineering Mechanics,
https://doi.org/10.1007/978-3-031-25401-7_10

were launched to attract more high school students to engineering and, in addition, to improve the quality of their education (e.g., ABET 2020; Engineers Australia 2019; Engineering Council UK 2021). Despite these efforts, many of today's engineering graduates are unable to meet the expectations of employers (Bloom and Saeki 2011; Markes 2006; Ramadi et al. 2016). Not surprisingly, the poor level of knowledge and skills exhibited by the graduates creates anxiety and confusion in the workplace (Jackson 2014; Kolmos and Holgaard 2019). One main cause for this is the passive manner of teaching and learning. These studies and others suggest that the continuing passive manner of lecturing, combined with students who learn only by listening and copying what the teacher presents, eliminates the opportunity for active inquiry and investigation of important connections to subjects taught in other classes. Thus, students end up with non-transversal knowledge and a lack of professional skills that leave them unprepared to successfully begin an engineer career.

The effectiveness of different pedagogical strategies has long been recognized, studied, and tested by philosophers, thinkers, and educators (i.e., Jean Piaget (1896–1980), Lev Vygotsky (1896–1934), John Dewey (1859–1952), Kolb (1939) or Benjamin Bloom (1913–1999)). The latest strategy of the effective teacher-learner method requires both students and teachers to think critically and actively inquire, rather than rely on lecture and passive memorization.

ABET 15, developed in the US by the important organization for accrediting an engineering program, encouraged the switch to new pedagogical models where teachers acting as coaches, that focused on improving the students' learning and professional capabilities. The responsibility for the sequence and the method of delivery of the required content is the responsibility of the main teacher. The preparedness of the students Despite the recognition that the students need to be equipped with better personal learning skills, most institutions are still "promoting" teacher centered learning, encouraged via the use of Likert-type Student Evaluations of Teaching (SETs), student survey questionnaires to assess the teaching quality of an instructor at the end of the semester. Tragically, the teacher is considered the only person responsible for the efficacy of the entire educational process and the important teaching evaluation, used for administrative decisions about the teacher, are qualitative, measuring how a student likes the teacher and the content of the class.

This one-sided qualitative assessment has several unfortunate consequences that have been documented by many researchers such as Carrell and West (2010), Johnson (2006), Krautmann and Sander (1999), McPherson (2006), McPherson et al. (2009), Weinberg et al. (2009). They were able to show that the result of a teacher evaluation is positively correlated with the grade that the student receives. One interesting study by Krautmann and Sander (1999) found that the significant factors correlating with the result of a teacher evaluation are: (a) grade, (b) student's expected grade, and (c) experience of the teacher. It is thus understandable that the outcomes of SETs are negatively correlated with the expected or actual final grade for the class. (Similar results by Thum 1986). The literature is clear on considering SETs a biased instrument that fails to properly consider the quality of the various dimensions of active, inquiry based, student-centered teaching. However, the the SETs could be used to encourage change recommended by ABET 15 and others (Weinber et al. 2009) by considering only qualitative questions that focus on assessing the excellence of both, students use of active learning methods and a teacher applying student-centered teaching strategies.

As was discussed earlier, the majority of higher-level engineering programs depend on quantitative SET-type questionnaires (Bernold 2007) especially since the older members of the faculty have figured out, via trial and error, how to achieve a satisfactory "final grade" from a student. This article presents a concept of education, with its related methods, for a multi-dimensional approach to create a student-centered environment of teaching. The educational evaluations serve also as tools that encourage pedagogical change. In fact, the multi-dimensional quantitative survey considers all factors of successful student-centered teaching and learning.

2 Background

2.1 Effect of Student-Centered Education on Teaching Evaluation

Piaget (1983) and Vygotsky (1978) are but two scientists in education who focused on proving the efficacy of alternatives to the standard pedagogical strategies to fit the different method of teaching. Central to one new teaching philosophy, named constructivism, is the understanding that students need to actively construct their knowledge. It is essential that they are "surrounded" by a supportive learning environment, are actively involved in the educational process led by a skilled teacher who serves as facilitators of learning. In the traditional chalk-and-talk learning process, the transmission of knowledge is expected to be done by the teachers who "deliver" knowledge that is taken up by inactive students. Later they have to demonstrate their understanding of the new material on quizzes and tests. Despite the many observations about the ineffectiveness of this teaching practice, studies such as one by Seldin (1993) found that over 90% of the university professors rely on lecturing. Furthermore, McKeachie (1997), noted that students are also not trained in rating teaching accurately, especially when it comes to the all-inclusive question "*Overall, was your instructor an effective teacher?*". One key cause for this is the ambiguity surrounding the term "*effective*" (McKeachie 1997). Similarly, conclusions of a small but focused study emphasized that: "…engineering sophomores perceive active student-centered teaching as greatly ineffective…" Naturally, switching to a constructivist method where students need to actively solve real world problems using effective learning skills, such as sketching and teamwork, to solve the problem while creating new knowledge. Equally important is the use of a teaching evaluation system that fits the new teaching philosophy. In order for this to work successfully, it is essential that the teacher stops supplying theoretical knowledge on a plate, takes on the role of a lead-coach who sets objectives, designs active learning exercises, asks relevant questions, organizes professional presentations, etc. In short, the teacher who mainly scribbles onto the black/whiteboard to be copied down, need to turn into a challenging but helpful coach. It goes without saying that this method of learning asks students to develop totally new and unfamiliar study skills.

2.2 *Survey of the Significance of Teacher Centered Education*

There exists no comprehensive study data about the spread and designs of teaching evaluations surveys used by different universities in the USA. To establish an only basic understanding, a small number of actual survey questionnaires were collected from 20 Engineering Schools in the US. A key objective was to appraise the use of evaluation factors that the students to which students have to answer at the end of the first semester. Not surprising, the result showed a lack of a cohesive "evaluation" strategy to gain and accurate and objective understanding of what the students perceive as high-quality teaching. Furthermore, the analysis of the survey data shows that the questions used are mirror the performance of a teacher centered teaching scheme. Table 1 provides an abbreviated summary of the results. Exceptionally "eyepopping" insights stem from comparing the conclusions with McKeachie (1997) statements about the deplorable situation of teaching evaluations in the US by creating quantitative survey questions about the effectiveness of the teacher while the students lack to recognize what effective teaching is.

While reviewing the survey questions to identify the six top ranked questions asked, several interesting observations can be made (a–c):

(a) Every question 1–14 focused on evaluating the only on the performance of the instructor.

Table 1 Partial List of the traditional teaching assessment questions used by 20 US University

Rank	Question	% of students responding
(1)	Instructor was the effectiveness of teacher	100 of responses
(2)	Instructor made clear presentation	80 of responses
(3)	Instructor was available when needed	60 of responses
⋮	⋮	⋮
(7)	Students learned a lot from teacher	50 of responses
⋮	⋮	⋮
(12)	Instructor increased the interest in the subject	40 of responses
⋮	⋮	⋮
(14)	Instructor's knowledge in the subject is perfect	35 of responses

(b) The most mentioned question was related to the overall effectiveness of the professor. Of course, this is exactly what McKeachie (1997) complained about.
(c) An important result if the fact that 40% of the students found that the teacher increased their interest in the subject. Various surveys showed that this issue is one of the key criteria resulting from constructivist-based evaluation.

3 Teaching and Learning in a Student's Center Learning Environment

Students to actively create their knowledge is the basis for implementing constructivism that originated from studies developed by Piaget (1972). Teaching–learning experiences must be built around a scaffolding that allows students to use their previous knowledge to acquire new information through a process (Piaget 1983). Complementarily, Vygotsky (1978) developed the theory of Sociocultural Constructivism that emphasizes that this scaffolding process is a social process generated through interaction and mediated by its cultural environment. Vygotsky (1978) declared that "knowledge emerges from people's interactions with their culture and community". That is, knowledge and learning is a social process in which the learner is an active agent. The relationship of how scaffolding takes place based on what we know, what we can learn on our own, and what we can learn with the help of others, is proposed by The Zone of Proximal development concept. In Vygotsky's theory, the concept of the zone of proximal development is defined as the distance between what individuals can accomplish alone and what they are able to accomplish when working with a peer.

Therefore, educational institutions must generate a learning environment which I considered not only teaching, but also learning and having tools that allow the development of learning. For their part, teachers will be a guide for students to build their own learning. Indeed, teaching and learning are complementary and dependent process that are mediate and nurturing through tools, rich environment, etc. Figure 1 provides a graphically sketch presents those three main elements that have been recognized as contributing to education in an integrated form.

As shown, teaching and learning are represented as "Ying and the Yang", constituting the core of the constructive educational integrated components. The third, also critical important, is the presented as a complete ring and should highlight the nurturing support of the environment such as the university, family and friends. The research university must facilitate inquiry in such contexts as the library, the laboratory, the computer, and the studio, with the expectation that senior learners, that is, professors, will be students' companions and guides." It is apparent that the university and the family have to function as psychological underpinning and the university provide the infrastructure to support the students (Boyer Commission 1998), the teachers are made responsible for coaching the individual student. Studying at an university, is in itself a profession requiring craftsmanship in such diverse areas as time management, test taking or metacognition (Bernold 2007).

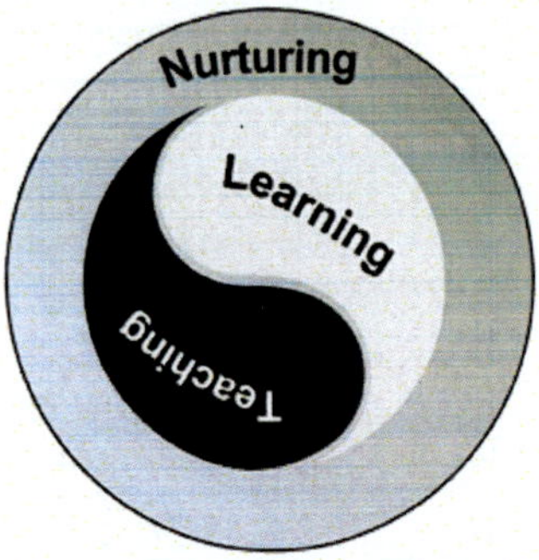

Fig. 1 A three component model of a holistic education in engineering

The following section introduces the basic architecture of a flexible and multi-dimensional educational assessment tool geared towards supporting constructive education.

3.1 Basic Elements of a Constructivist Course Evaluation Framework

As was discussed earlier, constructivist education depends on students and teachers to take on new roles. The teacher has to set up problems that raises the students' curiosity and interest (e.g., Tsunami Protection in Chile). It should be big enough that one student alone is unable to solve it but require multiple skills to find a solution to present to others. Different than in traditional teaching, students are encouraged, not forbidden, to work in teams. In fact, a constructivist teacher creates teams to work together, giving them different roles such as planer, recorder or information manager. It is apparent that working in teams is how work is done in the industry to solve problems even across continents. Off course the skill basked will have to change and so as the evaluation of the educational process.

The statistician and quality control expert, Edward Deming (1900–1993), was a first strong supporter of team work. In his many publications and video shows he makes clear that the path we are teaching and promoting is wrong. "*We have been taught by economists that competition will solve our problems. Actually, competition, we see now, is destructive. It would be better if everyone would work together as a system, with the aim for everybody to win.*" It is self-evident that working in teams fosters active and reflective learning. In effect, a group models working teams where every participant holds one piece of the puzzle. No one can do it all alone, but each person depends on the other group members."

Table 2 Categories of factors describing successful student-centered education

Environment	Learner/Student	Teacher
• Social network	• Engagement, group work	• Active teaching skills
• Supportive family	• Metacognition	• Teaching around the circle (Bloom)
• Excellent library	• Use of study methods	• Classroom management
	• Engineering sketching	

3.2 *Changes Needed to Move from Traditional to Inquisitive Learning*

A higher-level engineering student at a well-known school once complained that he was asked to read a book or text to prepare for class: "*How can you expect me to read something when I don't understand it?*" This attitude can also be found in Australia, where the student is considered a customer, a large amount of students complain to the administration when they are not first lectured to before being asked a question for discussion Of course, efficient reading and metacognition skills are essential for working to find information or to communicate what was found to the other team members. Without a doubt, the task to explain findings to the other team member is one of the most effective learning experiences.

Table 2 presents the main categories of essential performance factors to be considered by teachers and students working as a team.

Focusing on the component teaching, some categories could be seen like the traditional evaluations. But the focus and assessment are very different. As was we discussed before, the role of the instructor is more of a coach and provides guidance, and feedback to accompany the students learning. Additionally, the dynamics in the class are very different and the instructor needs to have sufficient management skills. I still remember when we were implementing PBL project in a classroom with freshman engineering students (see more details of the experience: Aizman et al. 2017; Díaz et al. 2017) and some instructors visiting the classroom. One of the instructors was surprised and asked us why the students are outside the classroom and everyone talking to each other without clear control of the class. I said, well, they are learning together and being where they feel more comfortable working on the project. These are some of the new skills that teachers must have in a student-centered class, which are shown in Table 3.

Regarding the environment categories, there are some that are given by the background of the students such as the social network and family support that are critical for success in higher education. In one of our investigations in progress, we have found that when knowing in detail students who have low family support, their main reason is not related to not having a "family present" but rather that none of their family members have the tools and don't know how to support their adolescent.

Table 3 Partial list of survey factors to assess student-centered-teaching

	Assessment factors	Indicators	
		Positive	Negative
E. Environment	Importance of education shared by others	Participation in study related activities	No extracurricular activities
	Access to adequate learning facilities	Library is available; access to internet; family support, house mentor (big-sister/brother)	Library closed on evenings/weekends. No internet. No family support
L. Learner/Student	Metacognition, individual learning preferences	Leverage personal, strengths to knowledge acquisition; ability to adapt approach to learning	No competence of self-assessment/regulation
	Effective use of methods	Daily/weekly planner, reading with method note taking; mentoring others (teaching) self-testing method	Procrastination; cramming before tests; focus on solving more example problems
	Personal health	Healthy diet, sufficient sleep, physical activities	Irrelevant to the life of the student; no connection to perceived real world
T. Teacher	Teaching skills	Considers different learning styles pedagogical expertise	
	Centered on students	Formative with encouragement. Accentuates relevance to students	Lot of lectures. Solving problems like on the test
	Classroom management	Interactive engagement of students. Discussions spaces	Lectures to passive students
	Fostering of professional skills	Professional presentations, formative writing assignments; development of portfolio	Focus on theory and examples

The University needs to respond to those problems with workshops designed to guide and support parents so that they provide adequate support to students. There are also other categories within the environment dimension that are the responsibility of the University and are found outside the classroom. For example, providing a library with an up-to-date, modern material that is easily available to students. Also, as discussed by Barbara Oakley et al. (2018), sport is vital to the learning process. The University must offer spaces to develop these activities.

The learner categories are associated with which students have the set of skills to achieve the expected learning. For example, Hampden-Thompson and Bennett (2013) confirmed our believe that students who get interested in the subject engage strongly: "Greater levels of student motivation, enjoyment, and future orientation towards science were found in classrooms where students reported that various measures of interaction, hands-on activities, and applications in science took place frequently". In addition, the fact that the student is in control of their learning requires that they have developed minimal metacognition skills for the effectiveness of the class. A teacher must recognize these skills of students, to develop their strategy and learning guidelines. Finally, we must recognize that not all students learn in the same way. Some students have received more than 12 years of a traditional education only and, therefore, their learning style does not may necessarily be the most compatible for an active class. In the next section, we show how those categories can be evaluated in an assessment instrument.

4 A Multi-dimensional Educational Quality Assessment Instrument

For the success of a student center education, it is crucial to have an instrument that allows evaluating and capturing all the categories discussed above: (a) Environment, (b) Learner and (c) Teacher. (Fig. 1) A first effort was made to build an instrument capable of handling each of these categories as a separate dimension. An example of clear and objective indicators concerning a quality of a student factors is the absence of procrastination or cramming. Table 3 shows the partially completed factor list.

Table 3 matrix is grouped into the three dimensions with factors. Important is the inclusion of the four-learning preference by Kolb (1971). Also considered are different motivation and commitment to learning. These factors will allow not only to comprehensively evaluate the learning experience, but also to be an input and feedback for teachers to improve their teaching practices.

5 Statistical Tools to Analysis of the Multidimensional Data

At the end of the semester the students are asked to fill out the multilayered questionnaire not necessarily to the "best" professor but to test much more interesting question. For example, exist there a correlation between the time when homework was handed in and the final grade in the class? Another interesting question could be is there a correlation between some learning styles of student and the type of group work that has been selected.

While a linear regression analysis of factor values is able rank their importance a multi regression determines the strengths of factors in influencing the educational effectiveness of the class. Both methods depend on factor models of an effective teacher, positive environment, and student actions (independent variables) to explain the ratings given by students/teacher about the quality of education (dependent variable). The correlation and p-value parameters of the regression allows us to determine which (or which) factors of a teacher's behavior or task selections are most effective in reaching which kind of students. A critical insight in the class work is provided by a factor analysis that can tell which factors most strongly create the correlations between them or creates the weakest and why? Carrell and West (2010), Johnson (2006), Krautmann and Sander (1999), McPherson (2006), and Weinberg et al. (2009) or maybe the correlation is associated with how much time or dedication the student is giving to the class. Other results could be that the student's opinion is biased by student-related aspects of the teacher's teaching skills.

It must be emphasized that this type of questionnaire needs to be designed in a quantitative manner where the student's answers need to be objective and verifiable. Additionally, we will be able to determine how the perception of the effectiveness of the class can be associated with gender or ethnicity. Considering that in engineering we have low participation of women, we need to determine how these minority groups perceive the effectiveness of the class and how we can help them to have a positive educational experience. By understanding what are the factors that minority groups value most in engineering, we can establish a systematic work that allows encourages more students from these groups to join engineering careers and also the students who are studying to have an effective degree.

Currently, universities have only opted to reduce the ethical conflict of "buy" evaluations (by giving higher grades to students) by applying the teaching evaluations a few weeks before the end of the semester. Of course, at that point the students have already received many partial notes that they will perfectly know what grade they will have. To eliminate this bias, using a multidimensional instrument will allow to establish a linear regression model to determine how class factors relate to student scores. For example, the students who receive the lowest grade will be those who have not participated in the class or have not. Additionally, one will be able to perform regressions between learning styles working in different groups. This will method provide numeric answers supported by a large group of students thus increasing the validity.

6 Conclusions

The common method of teaching in engineer worldwide is lecturing to inactive students by professors who are considered responsible for the quality of education that the students receive. Past and present research has shown that effective learning depends on the use of an appropriate pedagogical strategy where the students actively construct the new knowledge and skills rather than passively taking in information.

Research in engineering education makes clear that pedagogical improvements that bring theory to practice, require major changes in the educational process. The most significant theory underlying the new practices is constructivism which calls for moving from a teacher-centered to a student-centered teaching model. The paper contents that implementing such a model demands new learning skills, led by a teacher-coach who works with the students to understand and solve relevant and exciting problems.

Of course, adopting the constructivist philosophy is not simple or automatic. What are the incentives for investing time and energy for realizing the necessary changes? Since it is a crucial tool for the administration's teacher appraisals, a suitable approach is the use of a modified teaching evaluation process. As discussed in this text, many efforts to implement around the world. Unfortunately, the surveys used as instruments to evaluate teaching have not changed. Teacher evaluations continue to be based on opinions of students whose only model of learning has been the lecture and memorization format. Since the results of the teaching assessments may dictate the professors' future employment, this paper proposes to redesign the present system with one built on our suggested multi-dimensional model.

The constructivist teaching philosophy emphasizes the need for a holistic and multi-dimensional approach resembling a spiderweb of interwoven knowledge threads. Experimental work shows that real world problems stimulate students' interest while opening pathways linking to physics, trigonometry and other subjects necessary to solve the problem. Not surprisingly, the coined name for this method is Problem-Based Teaching.

As discussed in several sections of the paper, constructivist thinking, builds on three key elements: (1) A student who is a skilled learner, able to connect learning strategies, (2) a teacher who serves as a coach, able to successfully apply constructivist principles, and (3) an environment with a student friendly campus and supportive friends. Naturally, modelling comprehensively the three-element model, demands the use of a long list of factors. The authors discuss an effort to define multi-dimensional factors necessary for creating relevant questions to be used in a survey type evaluation. Of course, the results of such a survey will produce a lot of new data points that can be used for statistical analysis of student and teacher performance. A meaningful multi-factor evaluation would provide statistically sound information by which teachers could self-assess and improve.

ADDENDUM: CALL FOR COLLABORATION

The presented material is a work in progress with the goal to create a tool that is able to support the teachers and students who are willing to change. Due to the space limitation, we did not present all the work done so far. We would love to create a group of likeminded teachers across the globe to study how to expand the concept, to test and validate for special circumstances all with the goal to improving engineering in the long run.

References

ABET Engineering Accreditation Commission. Criteria for Accrediting Engineering Programs Effective for Reviews during the 2020–2021 Accreditation Cycle (2020)

Aizman, A., Bernold, L., Díaz, B., Olivares, A., Mora, C., Alarcón, H.: Introduction to engineering: a STEM approach. In: Proceeding of the 7th Research in Engineering Education. Research in Engineering Education Network, July 6–8, pp. 715–722. Bogota, Colombia (2017)

Bernold, L.E.: Evaluations for construction engineering and management: opportunity to move us forward. J. Constr. Eng. Manag. ASCE, **133**(2), 146–156 (2007)

Bernold, L., Ubilla, J., Alarcon, H., Garcia, G.: Coaching engineering learning skills to freshman class using STEAM and PBL. In: Proceeding of Sociedad Chilena de Educación en Ingeniería (SOCHEDI), 10–12 Oct, Vina de Mar. Chile (2018)

Bloom, B.S.: Taxonomy of educational objectives: the classification of educational goals. Cognitive Domain (1956)

Bloom, A., Saeki, H.: Employability and Skill Set of Newly Graduated Engineers in India. American Psychological Association (APA) (2011). https://doi.org/10.1037/e596532012-001

Boyer Commission on Educating Undergraduates in the Research University: Reinventing Undergraduate Education: A Blueprint for America's Research Universities. State University of New York at Stony Brook for the Carnegie Foundation for the Advancement of Teaching (1998)

Carrell, S.E., West, J.E.: Does professor quality matter? Evidence from random assignment of students to professors. J. Polit. Econ. **118**(3), 409–432 (2010)

Dewey, J.: Democracy and Education: An Introduction to the Philosophy of Education. Macmillan, NJ (1923)

Dewey, J.: Experience and Education. Collier-MacMillan Canada Ltd., Toronto (1938)

Díaz, B., Bernold, L., González, L.F., Aizman, A.: STE (A) M: Art, design and architecture in the formation of engineers at UTFSM. In: Proceeding of the 7th Research in Engineering Education. Research in Engineering Education Network, July 6–8, pp. 826–834. Bogota, Colombia (2017)

Engineering Council.: Standard for professional Engineering competence and Commitment (UK-SPEC). https://www.engc.org.uk/ukspec (2021)

Engineers Australia.: The Engineering profession, A Statistical Overview, Fourteenth Edution. https://www.engineersaustralia.org.au/sites/default/files/resources/Public%20Affairs/2019/The%20Engineering%20Profession%2C%20A%20Statistical%20Overview%2C%2014th%20edition%20-%2020190613b.pdf (2019)

Hampden-Thompson, G., Bennett, J.: Science teaching and learning activities and students' engagement in science. Int. J. Sci. Educ. **35**(8), 1325–1343 (2013)

Hossain, M., Robinson, G.M.: How to motivate US students to pursue STEM (science, technology, engineering and mathematics) careers. Online Submission, US-China Education Review A 4, p. 442–451 (2012)

Jackson, D.: Factors influencing job attainment in recent bachelor graduates. High. Educ. **68**(1), 135–153 (2014). https://doi.org/10.1007/s10734-013-9696-7

Johnson, V.E.: Grade Inflation: A Crisis in College Education. Springer Science & Business Media (2006)

Kolb, D.A.: Organizational Psychology. Englewood Cliffs, Prentice-Hall, NJ, USA (1971)

Kolmos, A., Holgaard, J.E.: Employability in engineering education: are engineering students ready for work? In: Christensen, S.H., Delahousse, B., Didier, C., Meganck, M., Murphy, M. (eds.) The Engineering-Business Nexus: Symbiosis, Tension and Co-Evolution, pp. 499–520. Springer International Publishing (2019). https://doi.org/10.1007/978-3-319-99636-3_22

Krautmann, A.C., Sander, W.: Grades and student evaluations of teachers. Econ. Educ. Rev. **18**(1), 59–63 (1999)

Markes, I.: A review of literature on employability skill needs in engineering. Null **31**(6), 637–650 (2006). https://doi.org/10.1080/03043790600911704

McKeachie, W.J.: Student ratings: the validity of use. Am. Psychol. **52**, 1218–1225 (1997)

McPherson, M.A., Jewell, R.T., Kim, M.: What determines student evaluation scores? A random effects analysis of undergraduate economics classes. East. Econ. J. **35**(1), 37–51 (2009). https://doi.org/10.1057/palgrave.eej.9050042

McPherson, M.A.: Determinants of how students evaluate teachers. J. Econ. Educ. **37**(1), 3–20 (2006). https://doi.org/10.3200/JECE.37.1.3-20

Montaigne.: 1580 On Schoolmasters' Learning. M. Screech, trans., Book I, p. 154 (1991)

Lane, E.S., Harris, S.E.: A new tool for measuring student behavioral engagement in large university classes. J. Coll. Sci. Teach. **44**(6), 83–91 (2015)

Loconte, J.: Bologna: Birthplace of the University. National Review (2021). https://www.nationalreview.com/2021/08/bologna-birthplace-of-the-university/

Oakley, B., Sejnowski, T., McConville, A.: Learning How to Learn: How to Succeed in School Without Spending all Your Time Studying; A Guide for Kids and Teens. Penguin (2018)

Olson, S., Riordan, D.G.: Engage to excel: producing one million additional college graduates with degrees in science, technology, engineering, and mathematics. Report to the president. Executive Office of the President (2012)

Piaget, J.: Intellectual evolution from adolescence to adulthood. Hum. Dev. 15, 1–12 (1972). http://dx.doi.org/10.1159/000271225

Piaget, J.: Piaget's Theory. In P. H. Mussen, & W. Kessen (eds.) Handbook of Child Psychology, vol. I History, Theory, and Methods (pp. 41–102). New York: John Wiley (1983)

Piaget and Vygotsky: theory and practice in early education. In: Good, T.L. (ed.) 21st Century Education: A Reference Handbook, vol. 1. Sage Publication (2008)

Ramadi, E., Ramadi, S., Nasr, K.: Engineering graduates' skill sets in the MENA region: a gap analysis of industry expectations and satisfaction. Eur. J. Eng. Educ. **41**(1), 34–52 (2016)

Seldin, P.: The use and abuse of student ratings of professors. Chronicle High. Educ. **39**(46), A40 (1993)

Simkin, J.: Education in the Roman Empire. Spartacus Educational Publisher, Ltd. (2020). www.spartacus-educational.com. Accessed 1 June 2022

Vygotsky, L.S., Cole, M.: Mind in society: Development of higher psychological processes. Harvard University Press (1978)

Weinberg, B.A., Hashimoto, M., Fleisher, B.M.: Evaluating teaching in higher education. J. Econ. Educ. **40**(3), 227–261 (2009)